The Self-Running Chicken Coop

ISBN: 978-1-962557-09-2

Table of contents

Why a Self-Running Chicken Coop?

Raising chickens can be one of the most rewarding backyard projects, but anyone who has owned chickens knows that it comes with daily tasks and occasional obstacles, or misfortunes. Whether it is cleaning out the coop, refilling water containers, or ensuring the chickens are fed, keeping chickens can feel like a chore. But what if there were a way to make your chicken-keeping experience nearly self-sufficient, reducing the amount of daily effort you put in while still maintaining a healthy and happy flock?

That is where the concept of a self-running chicken coop comes in. This book is designed to show you how to build a system that makes chicken keeping easy and manageable, even if you are busy or new to the process. By leveraging simple, effective strategies and smart designs, you will be able to create a chicken coop that takes care of many of the day-to-day tasks on its own, giving you more time to enjoy your chickens rather than just working for them.

The Challenge of Chicken Keeping

For first-time chicken owners or even seasoned pros, the routine can become repetitive. Early morning wake-up calls to let the chickens out, worrying about predators at night, and constantly ensuring they have enough food and water can feel like a full-time job. Then there is the less glamorous side—cleaning droppings, maintaining nesting boxes, or dealing with health issues.

While raising chickens is fulfilling, it does not have to come at the cost of your free time and peace of mind. With a few adjustments, you can automate much of the process and design a coop that virtually runs itself. This book aims to show you how to simplify your chicken-keeping routine with methods that work for both small backyard flocks and larger setups.

What Is a Self-Running Chicken Coop?

At its core, a self-running chicken coop is designed to reduce your involvement in the daily upkeep of your flock. While no system is truly 100% hands-off, this approach will help you automate essential tasks like feeding, watering, and cleaning. Imagine a coop where the doors open automatically at sunrise and close securely at sunset, where water is always fresh, and where feeding only needs your attention every week or two.

By building a coop that works with minimal human intervention, you can create a sustainable and efficient system that takes much of the burden off you, the chicken keeper, while ensuring your chickens have everything they need to thrive. We will look at strategies to:

Automate feeding and watering systems so your chickens always have access to clean food and water

Install predator-proofing solutions to keep your chickens safe day and night

Design the coop for easy cleaning and maintenance, reducing your time spent on messy tasks

Use the technology of automated doors and lighting to make the coop almost fully self-regulating

Why a Self-Running Coop Makes Sense

Whether you have few chickens in your backyard or a larger flock, a self-running coop makes life easier for you while keeping your chickens happier and healthier. By minimizing daily chores and focusing on long-term efficiency, you are creating a system that benefits both the birds and yourself.

Time Savings: One of the main benefits of a self-running system is the time saved from daily tasks. You will still need to check on your flock, but many of the more tedious chores, such as feeding and watering, can be automated.

Healthier Chickens: Automated systems mean your chickens are less likely to run out of food or water, which can sometimes happen if you are busy. Additionally, better ventilation and cleaning designs reduce the risk of disease.

Fewer Mistakes: When you are busy or tired, it is easy to forget to lock the coop at night or check if they have enough water. A self-running system takes care of those small, forgettable details for you.

Sustainability: Many of the ideas in this book focus on sustainable practices, like composting waste or using solar power, which reduces your environmental footprint and creates a more eco-friendly homestead.

Overview of What the Book Will Cover

This book will guide you through every aspect of building and maintaining a self-running chicken coop, from the initial planning stages to fine-tuning systems for efficiency. We will break down the process into simple, actionable steps that are easy to follow, even if you are completely new to chicken keeping or construction.

Here is a brief overview of what is to come:

Planning Your Coop: Before you build, you need a solid plan. We'll talk about choosing the right location, the different types of coops you can build, and how to ensure your setup works for your climate.

Building for Efficiency: The structure of your coop can make or break how easy it is to maintain. We will cover the best materials, designs, and layout for a coop that is not only durable but also low-maintenance.

Automating the Essentials: Learn how to set up automated systems for feeding, watering, lighting, and egg collection, as well as ways to predator-proof your coop for peace of mind.

Managing the Flock: We will discuss simple, effective ways to keep your chickens healthy and happy, including easy cleaning routines, biosecurity measures, and troubleshooting common issues.

Sustainable Practices: From composting chicken manure to using solar-powered systems, you will discover methods that make your coop eco-friendly and cost-effective.

By the time you have finished this book, you will be ready to build a chicken coop that practically runs itself—saving you time, energy, and hassle, while keeping your chickens in great shape.

The Benefits of Building a Self-Sustaining System

Raising chickens does not have to be a time-consuming job. With a little upfront planning and effort, you can build a coop that functions efficiently with minimal daily input. This means more time for the things you enjoy, whether that is gardening, spending time with family, or simply relaxing. And the best part? Your chickens will be healthier and more productive, too.

Through this book, you will gain the knowledge to design a system that fits your needs, regardless of the size of your flock or your level of experience. Whether you want to keep just a few chickens or expand into a larger operation, this guide will help you create a system that works for you.

Brief Benefits of a Self-Running Chicken Coop

Less time spent on chores: Automated feeding and watering systems free up your time.

Healthier chickens: Clean water, a well-ventilated coop, and stress-free environments help your chickens thrive.

More eggs: Chickens that are well cared for and live in optimal environments lay more eggs consistently.

Easier expansion: Once the system is in place, you can easily scale your flock without multiplying your workload.

Sustainability: By using eco-friendly solutions like rainwater harvesting, solar powered lighting, and composting waste, you reduce the environmental impact of your chicken-keeping hobby.

The Foundation of Success – Location

When building a chicken coop, many people focus on the structure itself: how big it will be, what materials to use, and how to keep the chickens happy. However, before you even begin construction, one of the most important decisions you will make is where to place your coop. A good location can save you hours of work in the future, prevent common problems like flooding or predator attacks, and ensure your chickens stay healthy and comfortable.

In this chapter, we will cover the key factors to consider when choosing a coop location, from weather exposure and drainage to predator protection and proximity to your home. Each element plays a role in creating a self-running chicken coop that simplifies maintenance and ensures the well-being of your flock.

Key Factors for Choosing the Right Location

Selecting the right spot for your chicken coop is not as simple as finding a sunny patch of land. There are several factors you need to consider to ensure long-term success. A well-thought-out location will not only make caring for your chickens easier, but it will also contribute to their health and safety.

Sunlight and Shade

Chickens thrive when they have access to both sunlight and shade. They need sunlight to help regulate their laying cycles and overall health, but too much exposure can lead to overheating, especially in warmer climates.

Morning Sunlight: Positioning your coop where it receives morning sunlight is ideal. This helps chickens start their day naturally and encourages egg production. Morning sunlight also helps dry out the coop, reducing moisture and the risk of mold and bacteria.

Shade in the Afternoon: Afternoon shade is equally important, especially during the hot summer months. Chickens are prone to heat stress, so placing your coop near trees or in a spot where it will be shaded in the afternoon can make a huge difference. If natural shade isn't available, consider creating some with a shade cloth or a roof extension.

Avoiding Constant Shade: While shade is essential in the heat, you do not want your coop in constant darkness. Too much shade can lead to dampness, mold, and lower egg production. Balance is key.

Wind Protection

Another crucial aspect of coop placement is protection from strong winds. Chickens do not do well in drafty environments, especially during the winter months when wind can cause cold stress and increase the risk of illness.

Natural Windbreaks: Try to place your coop near a natural windbreak like a line of trees, a fence, or even the side of your house. This will block strong winds and help keep the coop warmer in the winter without trapping too much heat in the summer.

Strategic Coop Orientation: Positioning the coop with the opening facing away from prevailing winds will also reduce the risk of cold drafts entering the coop. This is especially important if your area experiences frequent windstorms.

Elevation and Drainage – Keeping the Coop Dry

Water management is one of the most critical aspects of maintaining a clean, healthy, and low-maintenance chicken coop. Poor drainage can lead to soggy bedding, wet chickens, and the growth of harmful bacteria or mold. To avoid these problems, you need to think carefully about the elevation and drainage of your coop's location.

Choosing Higher Ground

Always place your coop on higher ground if possible. Low-lying areas are prone to flooding or becoming muddy during rainstorms, which can quickly turn your coop into a soggy mess. Chickens dislike wet conditions, and a damp coop can lead to health issues like respiratory infections.

Avoid Valleys and Dips: Avoid placing your coop in valleys, dips, or other areas where water tends to collect. Even slight slopes can direct water toward your coop if it is poorly positioned.

Artificial Elevation: If your property is mostly flat, consider raising the coop on a platform or adding extra fill to elevate it. This ensures water flows away from the coop rather than pooling underneath it.

Improving Drainage Around the Coop

Good drainage is essential for keeping the area around your coop dry and clean. Even in dry climates, rain or snowmelt can turn a poorly drained coop into a muddy disaster.

Gravel or Sand Base: Laying down a gravel or sand base under and around your coop can help improve drainage. These materials allow water to filter through rather than creating puddles.

Creating Drainage Trenches: If your coop is in an area prone to heavy rain, consider digging shallow drainage trenches around the perimeter of the coop to divert water away.

Slope for Runoff: When designing your coop, make sure the roof slopes away from the run and living area so that rainwater flows off quickly and does not pool near the chickens' living space.

Proximity to Your Home – Convenience Matters

How far should your chicken coop be from your house? The answer depends on how much interaction you expect to have with your chickens on a daily basis, but there are a few guidelines that will make your life easier.

Daily Access

If your coop is too far from your house, even simple tasks like collecting eggs or checking on your chickens can become inconvenient, especially in bad weather. On the other hand, placing the coop too close to your home can bring noise, smells, and pests right to your doorstep.

Ideal Distance: A good rule of thumb is to place your coop within easy walking distance from your home, but far enough away (about 25-50 feet) that odors and noise aren't an issue. This way, you can quickly check on your chickens without trekking across your property.

Water and Electricity Access: Consider whether you will need water or electricity near the coop. For instance, automated doors, lights, or heated waterers will require a power source. If your coop is close to the house, you can easily run an extension cord or water hose. If it is farther away, you may need to install a more permanent solution.

Security and Monitoring

The closer your coop is to your home, the easier it will be to keep an eye on your flock. Proximity to your home can also help deter predators like raccoons, foxes, and even hawks, which are less likely to approach when there is regular human activity nearby.

Visibility from the House: It is a good idea to place the coop in a location where you can easily see it from a window or your yard. This allows you to check on your chickens at a glance and keep an eye out for potential issues, such as predators or unexpected damage to the coop.

Predators and Protection – Keeping Your Flock Safe

Chickens are vulnerable to many predators, including raccoons, coyotes, foxes, hawks, and even neighborhood dogs. When choosing a location for your coop, one of your primary considerations should be how to protect your chickens from possible threats.

Understanding Local Predators

Before placing your coop, it is essential to understand the types of predators that are common in your area. Rural locations may face more threats from wild animals such as coyotes or raccoons, while suburban areas may have to deal with stray dogs or neighborhood cats.

Identify Vulnerabilities: Different predators require different types of protection. A coop near woods or fields may need stronger predator-proofing, such as buried wire or electric fencing, while a suburban coop may simply need to be enclosed securely to keep out neighborhood pets.

Predator-Proofing the Coop Location

Fencing and Barriers: Wherever you place your coop, ensure it is surrounded by sturdy fencing. Chicken wire or hardware cloth with small gaps can prevent many predators from digging or squeezing through. For extra security, bury the fencing a foot or more underground to deter digging animals.

Automatic Doors: If predators are a major concern in your area, consider installing an automatic coop door. These doors open at sunrise and close at sunset, ensuring your chickens are safely locked inside each night without you having to remember to do it manually.

Zoning Laws and Regulations

Before you finalize your coop location, it is essential to check local zoning laws and regulations regarding backyard chickens. Many cities and towns have specific rules about where you can place your coop and how far it needs to be from property lines or neighbors' homes.

Understanding Local Regulations

Setback Requirements: Many areas require that coops be placed a certain distance from property lines, streets, or neighboring homes. Failing to comply with these regulations can lead to fines or being forced to move your coop after it is already built.

Permits and Restrictions: In some locations, you may need a permit to build a chicken coop, especially if it is a larger structure. Additionally, certain areas restrict the number of chickens you can keep or ban roosters due to noise concerns.

Neighbor Relations

Even if your local laws are lenient, it is always a good idea to consider your neighbors when choosing a coop location. No one wants to be woken up by a noisy flock at 5:00 in the morning or deal with odors wafting into their yard. Placing your coop a respectful distance away from property lines can help prevent conflicts and maintain friendly relationships.

Coop Types and Their Impact on Maintenance

Choosing the right type of coop is one of the most important decisions you will make when setting up a self-running system. The size, style, and functionality of your coop will directly affect how much time you spend maintaining it. Some designs require daily upkeep, while others can be designed for long-term ease of use with automated systems built right in.

In this chapter, we will explore the various types of coops, from fixed, permanent structures to mobile coops that allow you to move your chickens around. We will compare different designs based on ease of maintenance, automation potential, and the unique needs of your flock. By the end of this chapter, you should have a clear idea of what coop style will best fit your goals for minimal maintenance and maximum efficiency.

Fixed vs. Mobile Coops

The first decision you will need to make is whether you want a fixed or mobile coop. Each option has its pros and cons, especially when it comes to long-term maintenance. Let's break down what to expect from each.

Fixed Coops – The Stable Solution

A fixed coop is a stationary structure that remains in one place year-round. These coops tend to be larger and more durable than mobile coops, making them ideal for those who plan to keep chickens in the same spot and want the ability to install more permanent features like automatic doors, electricity, and water lines.

Maintenance Considerations: Fixed coops can be designed with self-running systems in mind. Because they do not move, you can easily install automated systems like feeders, waterers, and lighting. However, one downside is that over time, the area around the coop can become worn down, leading to issues like mud buildup or a lack of fresh grass for your chickens to forage on. Managing drainage and keeping the area clean requires more planning.

Best for: Large flocks or those who prefer a stable, permanent structure. A fixed coop is also ideal if you want to install more advanced automation systems or use a deep litter method that benefits from a stationary setup.

Mobile Coops – Flexibility with Less Wear and Tear

Mobile coops, also known as "chicken tractors," are designed to be moved from place to place. These coops are typically smaller, allowing you to rotate your flock to fresh ground, which helps prevent the buildup of manure in one spot and keeps your chickens on fresh grass.

Maintenance Considerations: Mobile coops can be a great option for reducing some types of maintenance. Because you can move them regularly, you won't have to worry about cleaning out heavily soiled areas as often. The chickens also help maintain the grass as they forage. However, mobile coops may require more frequent checks since they tend to be lighter and may need more predator-proofing.

Best for: Smaller flocks or those who want to reduce the amount of cleaning by rotating their chickens to new areas. Mobile coops are also great for those with larger properties who want to keep their grass healthy.

Size and Space Requirements – Finding the Right Balance

One of the most common mistakes new chicken owners make is either building a coop that's too small or going overboard and creating a space that's much larger than necessary. The size of your coop has a significant impact on the ease of maintenance, the health of your flock, and your overall ability to automate systems.

Calculating Coop Size Based on Flock Size

As a general rule, chickens need about 3 to 4 square feet per bird inside the coop and 8 to 10 square feet per bird in the outdoor run. Too little space can lead to overcrowding, which results in stress, fighting, and health problems. On the other hand, too much unused space can make heating and cleaning more difficult.

Optimal Size for Automation: If you plan to install automatic systems, consider how much room you will need for these features. For instance, an automatic feeder and waterer may require additional space. However, having a well-sized coop prevents the chickens from clustering in one area and reduces the amount of bedding needed, which simplifies cleaning.

Coop Height and Ventilation

Height is another important factor. Chickens do not need much vertical space, but having a coop that's tall enough for you to stand in will make cleaning and maintenance much easier. Tall coops also allow for better ventilation, which is essential for maintaining good air quality and reducing moisture buildup.

Ventilation Considerations: Proper airflow helps prevent respiratory issues and reduces the need for frequent deep cleaning. Higher ceilings and good vent placement can create a natural flow of air, helping your coop stay fresh with minimal effort. A well-ventilated coop will also reduce moisture, which cuts down on the risk of mold and bacteria growth.

Automated vs. Manual Coops – Which Is Right for You?

One of the main goals of a self-running chicken coop is automation. By incorporating automated systems, you can significantly reduce the amount of hands-on time you need to spend on tasks like feeding, watering, and opening or closing the coop. However, not every coop type is suited for full automation, and some setups may require a hybrid approach that balances automation with manual effort.

Fully Automated Coops

A fully automated coop includes systems that handle essential tasks like feeding, watering, and providing light, often using electricity or solar power. Automated doors can open at dawn and close at dusk, ensuring your chickens are safely inside at night without you having to intervene. This setup is ideal for those who want to minimize daily involvement.

Best Coop Types: Fixed coops are typically best for full automation since they provide a stable location for electrical setups or solar panels. These coops can be equipped with all the necessary technology, making them virtually self-sufficient.

Maintenance Considerations: Fully automated systems require some upfront investment and occasional maintenance, such as replacing batteries or cleaning the components. However, the long-term benefits far outweigh the occasional check-ins.

Semi-Automated Coops

If you are not ready to go fully automated, a semi-automated coop can still save you significant time while maintaining some manual control. For example, you might install an automatic waterer and feeder but choose to open and close the coop manually or clean it regularly.

Best Coop Types: Semi-automated systems work well with both mobile and fixed coops. Mobile coops can incorporate gravity-fed systems for feeding and watering, reducing the need for electricity, while fixed coops can use simpler timers and solar setups without being fully automated.

Maintenance Considerations: Semi-automated coops give you more control over certain tasks, which may be ideal if you enjoy interacting with your flock daily. However, the trade-off is that you will need to perform these tasks more regularly compared to a fully automated system.

Comparing Popular Coop Styles

There are several popular coop designs, each with unique benefits and challenges. Choosing the right design for your flock depends on your goals, available space, and how much maintenance you are willing to commit to.

The Classic Walk-In Coop

A walk-in coop is a full-sized structure that allows you to stand and move around inside. These coops are often designed to house larger flocks and include features like nesting boxes, roosting bars, and built-in storage for feed and supplies.

Pros: Easy to clean, ample space for installing automated systems, and good ventilation.

Cons: Requires more space and materials to build, and typically more expensive upfront.

The A-Frame Coop

A-frame coops are compact, triangular structures that offer a simple, space-saving design. These coops are often mobile, allowing you to move them to fresh grass as needed. They are typically designed for smaller flocks.

Pros: Lightweight, mobile, and easy to construct. Ideal for small spaces and beginner chicken keepers.
Cons: Limited space for automated systems, can be harder to clean, and may not provide as much protection in extreme weather.

The Tractor Coop (Mobile Coop)

A tractor coop, or mobile coop, is designed to be moved regularly. These coops are usually smaller and feature wheels or handles for easy mobility. The idea is to give chickens access to fresh ground while reducing the need to clean heavily soiled areas.

Pros: Reduces the need for cleaning by rotating the coop, great for keeping grass healthy, and reduces odor buildup.
Cons: Limited space for large flocks or automated systems, can be prone to tipping or predator attacks if not securely built.

Features of a Self-Running Coop

No matter which type of coop you choose, certain features will help make your system more self-sufficient. By incorporating these elements, you can drastically reduce the time and effort needed to maintain your flock.

Automated Feeders and Waterers

Investing in gravity-fed or automated feeders and waterers is one of the simplest ways to reduce daily maintenance. These systems ensure your chickens always have access to clean food and water without you needing to refill them constantly.

Gravity-Feed Systems: Ideal for smaller flocks, these systems can hold several days' worth of feed and water, cutting down on the number of trips you need to make to the coop.

Predator-Proofing Essentials

Regardless of the type of coop you choose, predator-proofing is non-negotiable. A self-running system isn't very effective if your flock is constantly at risk. Secure fencing, automatic doors, and reinforced windows are all essential for keeping predators out.

The Deep Litter Method

If you are looking to reduce cleaning time, the deep litter method is a popular strategy. Instead of frequently cleaning out the coop, you allow bedding material like wood shavings or straw to accumulate and compost naturally over time. This method not only reduces how often you need to clean but also provides heat in the winter as the material decomposes.

By understanding the pros and cons of different coop types and selecting features that support a self-sufficient system, you can create a setup that requires minimal day-to-day involvement. Whether you opt for a fixed, mobile, or hybrid system, the key to success lies in thoughtful planning and a commitment to reducing manual labor without compromising on chicken welfare.

Climate Considerations – Adapting Your Coop to Your Region

Designing for Hot Climates

If you live in an area where summers are long and hot, keeping your chickens cool is a top priority. Chickens are highly susceptible to heat stress, and temperatures that climb above 90°F (32°C) can put their health and productivity at risk. By designing your coop with hot weather in mind, you can create a comfortable space that minimizes the need for hands-on intervention during heat waves.

Ventilation is Key

The most critical element of a hot-weather coop is ventilation. Good airflow helps cool the coop and prevents the buildup of humidity, which can cause respiratory problems and mold growth.

Windows and Vents: Make sure your coop has plenty of windows and vents, particularly on the upper part of the structure where hot air can escape. Windows should be covered with predator-proof mesh, allowing air to flow freely without risking the safety of your flock.

Cross-Ventilation: Position vents on opposite sides of the coop to encourage cross-ventilation. This will allow cooler air to flow in while pushing hot air out. Consider adding adjustable vents that can be closed or opened depending on the temperature.

Roof Venting: If possible, include ridge vents along the roofline. Heat naturally rises, and roof vents will allow it to escape easily, preventing the interior of the coop from becoming stifling.

Shade and Insulation for Heat Reduction

Even with good ventilation, chickens will still struggle in extreme heat if the coop is in direct sunlight all day. Providing shade and insulating the coop can help keep temperatures inside at a comfortable level.

Natural Shade: Place your coop in a location that receives natural shade from trees, bushes, or nearby buildings. If natural shade isn't available, consider building a shade structure or using shade cloth to cover the roof and sides of the coop during the hottest parts of the day.

Insulation: Insulation isn't just for cold climates; it can also help keep a coop cool in hot weather by preventing the sun's heat from penetrating the walls and roof. Use reflective roofing materials or light-colored paint to reflect heat, and insulate the walls with eco-friendly materials like straw bales, which work well in both hot and cold conditions.

Cooling Strategies for Extreme Heat

In regions with extreme temperatures, you might need to incorporate active cooling strategies to keep your chickens comfortable.

Misting Systems: Installing a misting system around the run or on the outside of the coop can help lower the surrounding temperature. Chickens will often seek out misted areas to cool down. Make sure to only use these systems outside the coop to avoid excess moisture buildup indoors.

Frozen Water Bottles: Placing frozen water bottles or ice blocks in the coop can give your chickens a place to cool down. They'll sit next to or on top of the bottles to lower their body temperature.

Fans: While more manual, small solar-powered fans can be a useful addition in very hot climates. Place them in a way that does not create direct drafts on the chickens but helps circulate air through the coop.

Adapting for Cold Climates

On the opposite end of the spectrum, if you live in an area where winters are harsh, keeping your chickens warm and healthy during freezing temperatures is the biggest challenge. Proper insulation, protection from the wind, and ensuring the coop does not become too damp or humid are all essential for winter-proofing your setup.

Insulating the Coop for Cold Weather

Cold weather presents a different set of challenges, and insulation becomes a vital part of your coop's design. Keeping the interior warm without relying on electrical heaters or constant human intervention can save you time and money.

Wall and Roof Insulation: Insulate the walls, floor, and roof of the coop to retain warmth. Use materials like foam board, fiberglass, or even natural insulators like straw bales. Avoid materials that absorb moisture, as they can lead to mold and rot over time.

Deep Litter Method: One of the easiest ways to maintain warmth in a cold-weather coop is by using the deep litter method. As bedding material like straw or wood shavings builds up, it begins to compost naturally, generating heat. This method not only provides warmth but also reduces the frequency of coop cleaning in winter.

Preventing Drafts While Maintaining Ventilation

While keeping the coop warm is essential, it is equally important to ensure good ventilation to prevent moisture buildup, which can lead to frostbite or respiratory issues. Striking the right balance between insulation and ventilation is key.

Sealing Gaps: Close any gaps or cracks in the coop walls and doors that could allow cold drafts to enter. Make sure the windows and doors close securely, and consider using weatherstripping or caulking to seal any openings.

Upper Ventilation: In cold climates, place ventilation openings near the roofline. This allows moisture to escape without creating drafts near where your chickens roost. Proper ventilation helps to reduce humidity, preventing frost from forming inside the coop.

Providing Safe Heat Sources (If Necessary)

In extremely cold regions, you may need to add supplemental heat to the coop. However, it is important to do this safely to avoid fire hazards and not overheat the chickens.

Heated Perches: Rather than heating the entire coop, consider using heated perches. These can keep your chickens' feet warm, which is often the part of their body most susceptible to frostbite.

Heat Lamps and Panels: If you do use a heat source, avoid traditional heat lamps, which can be a fire hazard if knocked over. Instead, use radiant heat panels or heaters designed specifically for chicken coops. These provide safe, consistent warmth without overheating the entire coop.

Managing Humidity and Moisture

Moisture is one of the most common problems in coops, regardless of whether you live in a hot or cold climate. Excessive moisture leads to issues like respiratory infections, mold, and bedding that quickly becomes soiled. Here's how to manage moisture year-round.

Controlling Humidity in the Coop

Humidity inside the coop can quickly become a problem, particularly in humid climates or during rainy seasons. High humidity contributes to damp bedding and poor air quality, both of which can affect your chickens' health.

Ventilation Solutions: As we've mentioned, ventilation is crucial in all climates, but it is especially important in humid regions. In areas where humidity is high, ensure your coop has plenty of ventilation to allow excess moisture to escape. Use roof vents, windows, and fans if necessary to promote airflow.

Regular Bedding Changes: In very humid climates, you may need to change bedding more frequently to prevent it from becoming soggy and compacted. Even in colder climates, wet bedding can freeze, creating uncomfortable conditions for your chickens. Use materials like straw or wood shavings that absorb moisture without becoming waterlogged.

Preventing Water Buildup

Water inside the coop, whether from rain, spilled waterers, or condensation, can lead to health issues and make cleaning more difficult.

Waterproofing the Coop: Make sure your coop is designed to prevent water from entering. Sloped roofs help rainwater run off, while raised coops prevent groundwater from seeping in. If your area is prone to flooding, ensure the coop is on high ground or build a drainage system around the perimeter.

Spill-Proof Waterers: Use spill-proof or hanging waterers to prevent chickens from knocking over their water supply. This reduces the amount of moisture in the coop and keeps bedding dry. In cold climates, you may also want to consider heated waterers to prevent freezing.

Seasonal Maintenance – Preparing for Year-Round Changes

Every climate has seasonal changes, and your coop should be adaptable to handle the shift from one season to the next. Whether you are preparing for summer heat, winter cold, or the rainy seasons in between, having a seasonal maintenance routine will help your coop run smoothly.

Summer Prep: Cooling and Ventilation

As temperatures start to rise, prepare your coop for summer by checking ventilation systems, adding shade, and removing any insulation that might trap heat.

Check Vents and Fans: Ensure that all ventilation systems are working properly, and consider adding extra shade to the run area. If your coop uses solar-powered fans, ensure they are running efficiently to keep air flowing.

Hydration and Cooling: Double-check that water systems are functioning properly, as chickens will drink more during hot weather. If you use misting systems or frozen water bottles for cooling, prepare these early in the season.

Winter Prep: Insulation and Frost Prevention

Before winter hits, you will need to inspect the coop for drafts, ensure it is well-insulated, and get any heating systems ready.

Check Seals: Inspect the coop for any gaps or cracks where cold air could enter, and seal them with weatherstripping or caulk. Make sure the insulation is intact, especially in the walls and roof.

Winterize Water Systems: If you use heated waterers or other freeze-prevention systems, test them before the cold weather arrives to ensure they're functioning properly.

Wet Season Prep: Drainage and Flood Prevention

Rainy seasons bring their own challenges. Ensuring your coop is protected from water and mud will reduce the need for frequent cleaning and keep your chickens comfortable.

Check Roof and Drainage: Inspect the roof for leaks and make sure water drains away from the coop. If you notice pooling around the coop, add gravel or a drainage trench to divert water.

Prepare for Mud: If your chickens' run tends to get muddy during wet weather, consider adding straw or sand to help soak up excess moisture and keep the ground drier.

Climate-Specific Challenges and Solutions

Every region presents unique challenges when it comes to managing a chicken coop. Here's how to handle a few specific regional conditions to create a more efficient, self-running coop system.

Hot and Arid Climates

In dry, hot regions like deserts, keeping chickens cool without relying on excessive water usage is the main concern. Use materials that reflect heat, provide plenty of shade, and focus on maximizing ventilation to prevent overheating.

Cold and Snowy Climates

In snowy regions, make sure your coop is elevated to avoid snow piling up around the structure. Plan for snow removal, and install heated waterers to prevent freezing. Consider using clear plastic panels or greenhouse materials over the run to allow sunlight while blocking snow and wind.

Humid and Rainy Climates

In humid, rainy areas, focus on waterproofing and ventilation. Ensure your coop has a sloped roof, good drainage, and enough airflow to prevent moisture buildup. Consider using sand in the run to help absorb water and reduce mud.

Designing your chicken coop to suit your climate is a key part of making your system self-sufficient. By creating a setup that adapts to extreme heat, cold, or moisture, you will not only reduce the amount of time you spend on maintenance, but also keep your flock healthy and comfortable in all seasons.

Material Matters – Choosing Building Supplies

When building a self-running chicken coop, one of the most crucial decisions you will make is selecting the right materials. The materials you choose will impact the coop's durability, how easy it is to clean and maintain, and ultimately the health and safety of your chickens. Whether you are building a coop from scratch or upgrading an existing structure, this chapter will guide you through the best materials for creating a durable, low-maintenance, and chicken-friendly environment.

We'll cover various materials for different parts of the coop—such as walls, flooring, roofing, and the run—along with their pros and cons. By the end of this chapter, you will be able to choose the right building supplies for your climate, budget, and maintenance goals.

Best Materials for Coop Walls

The walls of your coop serve as the primary barrier between your chickens and the outside environment. They need to be sturdy enough to protect your flock from predators and the elements, while also being easy to clean and maintain. The choice of material will depend on your climate, your budget, and how much insulation you need to provide.

Wood – A Classic and Versatile Choice

Wood is one of the most popular materials for building chicken coops, and for good reason. It is readily available, easy to work with, and provides excellent insulation, which is important in both hot and cold climates.

Pros: Wood is sturdy, provides natural insulation, and is relatively easy to customize. It also looks more natural and blends in well with garden environments.

Cons: Wood can rot if exposed to moisture over time, especially if it is untreated. It is also more susceptible to pests like termites, and it will require occasional maintenance, such as repainting or sealing.

Best for: Wooden coops are great for most climates, especially if you are willing to maintain them. If you live in a particularly wet region, consider treating the wood with a weatherproof sealant.

Metal – Durability with Minimal Maintenance

Metal coops, often made from galvanized steel or aluminum, are highly durable and require very little maintenance. They are resistant to rot, pests, and harsh weather conditions, making them an excellent choice for areas with extreme weather.

Pros: Metal is extremely durable, resistant to weather, and requires minimal maintenance. It won't rot or warp over time, making it ideal for long-term use.

Cons: Metal can become extremely hot in the summer or cold in the winter if not properly insulated. It also does not provide the same aesthetic appeal as wood, and it can be more expensive upfront.

Best for: Metal coops are ideal for areas with high moisture or extreme weather, where durability is a top priority. However, you will need to ensure proper insulation and ventilation to prevent overheating or cold stress.

Plastic – Lightweight and Easy to Clean

Plastic coops, often made from heavy-duty molded plastic panels, are gaining popularity for their low maintenance requirements. These coops are lightweight, weatherproof, and resistant to rot, making them a practical option for many chicken keepers.

Pros: Plastic is easy to clean, weatherproof, and resistant to pests. It is also lightweight, which makes it easier to move if you need a mobile coop.

Cons: Plastic does not provide much insulation, so it may not be ideal for extreme climates. It can also crack or fade over time if exposed to prolonged sunlight.

Best for: Plastic coops work well in moderate climates where temperature extremes aren't a major concern. They are also great for those who want a low-maintenance, easy-to-clean setup.

Flooring Materials – Cleanliness and Durability

The flooring of your coop is another critical decision. It needs to be strong enough to support the weight of your chickens, easy to clean, and resistant to moisture and pests. A well-chosen floor can reduce the amount of time you spend cleaning and ensure a healthier environment for your flock.

Wood Flooring

Wood flooring is a common choice, especially in traditional wooden coops. It is relatively inexpensive and can be treated to last a long time. However, wood floors can absorb moisture, leading to rot and the potential for mold or bacteria buildup.

Pros: Wood is inexpensive, readily available, and provides some natural insulation.

Cons: Wood absorbs moisture, which can lead to rot and the buildup of bacteria if not cleaned regularly. you will need to treat the wood to protect it from moisture and pests.

Best for: Wooden floors are best for dry climates where moisture isn't a constant issue, or when paired with the deep litter method, which helps to manage moisture levels.

Concrete Flooring

Concrete is one of the most durable flooring materials you can choose for a chicken coop. It is impenetrable to predators, easy to clean, and resistant to moisture. However, it does not provide any insulation, so you will need to add bedding to keep your chickens comfortable.

Pros: Concrete is easy to clean, pest-proof, and moisture-resistant. It also provides a solid foundation that will last for years.

Cons: Concrete can be cold and uncomfortable for chickens without adequate bedding. It also requires a solid foundation, which can be expensive and difficult to install.

Best for: Concrete floors are ideal for permanent, fixed coops where long-term durability and ease of cleaning are the top priorities. They are particularly useful in areas prone to heavy rain or predators.

Dirt or Gravel Flooring

Dirt or gravel flooring is often used in mobile or more rustic coops. It is inexpensive and allows chickens to scratch and dig, which is great for their natural behaviors. However, it is harder to keep clean, and dirt floors can become muddy in wet weather, leading to hygiene issues.

Pros: Dirt and gravel floors allow chickens to forage naturally and are inexpensive to install. Gravel helps with drainage in wet conditions.

Cons: Dirt floors are harder to clean and can become muddy and messy, particularly in wet climates. They also offer little protection against predators that can dig into the coop.

Best for: Mobile coops or areas with good drainage. If using dirt or gravel, ensure there is predator-proofing in place, such as buried hardware cloth to prevent animals from digging under the coop.

Roofing Materials – Weather Protection and Durability

The roof of your coop will need to withstand the elements—rain, snow, heat, and wind. Choosing the right roofing material will ensure your coop lasts longer and requires less maintenance, while keeping your chickens dry and comfortable.

Metal Roofing

Metal roofing is one of the most durable and weather-resistant options available. It can withstand heavy rain, snow, and wind, and is resistant to rot and pests. However, it can also get very hot in the summer, so proper ventilation is essential.

Pros: Metal is highly durable, long-lasting, and requires very little maintenance. It is excellent for wet climates or areas with heavy snowfall.

Cons: Metal can become very hot in the sun, which can overheat the coop if there isn't proper insulation. It is also noisier during rainstorms, which may stress some chickens.

Best for: Areas with heavy rain or snow where durability and weather resistance are top priorities. Be sure to include adequate ventilation to prevent heat buildup.

Asphalt Shingles

Asphalt shingles are a common choice for chicken coops, offering a balance of durability, affordability, and ease of installation. They are weather-resistant and can last for many years with minimal maintenance.

Pros: Asphalt shingles are affordable, relatively easy to install, and offer good protection from the elements. They also provide some insulation, helping to regulate the temperature inside the coop.

Cons: Asphalt shingles aren't as durable as metal and may need to be replaced over time. They also add some weight to the structure, which could be an issue for smaller or mobile coops.

Best for: Most climates, particularly those with moderate rain or snow. Asphalt shingles are a good option for permanent coops that require weatherproofing without the higher cost of metal roofing.

Plastic or Polycarbonate Roofing

Plastic or polycarbonate roofing panels are lightweight, affordable, and often translucent, allowing natural light into the coop. They are durable and easy to install, making them a popular choice for backyard chicken keepers.

Pros: Plastic roofing is lightweight, easy to install, and lets in natural light, which can help with egg production. It is also waterproof and resistant to rot and rust.

Cons: Plastic roofing does not offer much insulation, and it can become brittle over time with exposure to sunlight. It may also warp or discolor if not properly maintained.

Best for: Areas with moderate weather, where light is needed inside the coop. Plastic roofing is ideal for small coops or runs, where affordability and easy installation are key.

Materials for the Run – Fencing and Ground Cover

The chicken run is an essential part of your coop's design, as it provides your chickens with outdoor space to forage and exercise. The materials you choose for the run will impact how easy it is to maintain and how well it protects your flock from predators.

Fencing Materials for the Run

Fencing is the most important element in keeping your chickens safe

from predators. The right materials will depend on the type of predators in your area and the size of your flock.

Hardware Cloth: Hardware cloth with small, ½-inch mesh is the most predator-proof fencing option. It is strong enough to keep out even the smallest predators, like weasels and snakes, and it can't be easily chewed through or bent.

Pros: Extremely durable and predator-proof. It can be buried around the perimeter to prevent digging animals from getting inside.

Cons: Hardware cloth is more expensive than standard chicken wire and can be harder to work with due to its rigidity.

Chicken Wire: While traditional chicken wire is cheap and easy to install, it is not as secure as hardware cloth. It is best used for containing chickens rather than keeping out predators, as many animals can chew through it or squeeze through larger openings.

Pros: Inexpensive and easy to install. Ideal for temporary runs or areas with low predator pressure.

Cons: Not predator-proof. Larger animals like raccoons or dogs can easily tear through chicken wire, and smaller predators can slip through larger gaps.

Ground Cover for the Run

The ground cover inside the run should be chosen based on your climate and how often you plan to clean or move the coop. Good ground cover will reduce mud, keep chickens dry, and prevent health issues related to dampness or dirt buildup.

Sand: Sand is a popular ground cover choice because it provides excellent drainage and stays relatively clean. Chickens love to dust-bathe in it, and it is easy to rake or sift clean.

Pros: Drains well, reduces odor, and stays dry in wet weather. It is also easy to maintain with regular raking.

Cons: Sand needs to be replaced periodically, and in very hot climates, it can become too warm for chickens to stand on.

Straw or Hay: Straw is another common choice, offering chickens a natural material to scratch and dig through. However, it can become soggy in wet weather, leading to mold and bacteria buildup if not changed regularly.

Pros: Inexpensive and easy to spread. Provides chickens with material to scratch through.

Cons: Needs frequent changing, especially in wet climates. Can harbor pests and bacteria if left too long.

Cost-Effective Solutions for Long-Term Maintenance

Building a coop from high-quality materials does not always have to break the bank. There are several cost-effective options that can provide long-term durability while keeping your budget in check.

Recycled or Repurposed Materials

Using recycled materials is a great way to save money while building an eco-friendly coop. Many items, such as old wood, pallets, or metal, can be repurposed to create a sturdy and attractive coop.

Pros: Cost-effective and environmentally friendly. Using recycled materials can also give your coop a unique, rustic look.

Cons: Recycled materials may require more work to prepare, such as sanding or weatherproofing, and they may not last as long as new materials.

Eco-Friendly and Sustainable Options

If you are concerned about the environmental impact of your coop, there are plenty of sustainable materials to choose from. Bamboo, reclaimed wood, and other eco-friendly options can help reduce your environmental footprint while still providing a durable structure.

Pros: Reduces environmental impact and often comes from renewable sources. Eco-friendly materials are also non-toxic, which is better for your chickens' health.

Cons: Sustainable materials can sometimes be more expensive, and they may require additional maintenance or treatment to ensure durability.

By choosing the right materials for your coop, you can create a structure that's durable, easy to clean, and comfortable for your chickens. Whether you opt for wood, metal, plastic, or a combination of materials, the goal is to build a coop that minimizes long-term maintenance while ensuring the safety and health of your flock.

Designing a Coop for Easy Cleaning and Maintenance

Cleaning and maintaining a chicken coop can be one of the most time-consuming aspects of keeping chickens. However, with the right design features, you can significantly reduce the amount of time and effort needed to keep your coop clean, dry, and healthy for your flock.

In this chapter, we will explore how to design a coop with easy access for cleaning, the benefits of the deep litter method, how to manage waste and drainage effectively, and the best tools and techniques for maintaining your coop with minimal hassle.

Layout and Design for Easy Access

One of the simplest ways to make cleaning and maintenance easier is by designing a coop that provides easy access to all areas. When you can reach every corner of your coop without much hassle, you will find that regular cleaning becomes less of a chore.

Walk-In vs. Low-Profile Coops

When it comes to accessibility, a walk-in coop can make a world of difference. These coops are tall enough for you to enter and move around comfortably, making it easier to clean, refill feeders and waterers, and collect eggs. However, smaller, low-profile coops can also be designed with easy access in mind.

Walk-In Coops: These coops allow you to easily reach all areas for cleaning and maintenance. The added height means you can store feed and tools inside, and chickens have more room to roost at different heights. However, walk-in coops take up more space and require more materials to build.

Low-Profile Coops: If you are working with a smaller coop, design it with access doors on all sides so you can reach the interior without having to crawl in. Consider removable panels or large doors that swing open to allow for easy cleaning.

Removable or Hinged Panels

A great way to simplify cleaning is by incorporating removable or hinged panels into your coop design. These panels allow you to quickly access the inside of the coop for spot cleaning or deep cleaning sessions.

Removable Trays or Floors: A removable floor or tray under the roosting bars can catch droppings and make cleaning easier. You can simply slide out the tray, dump the waste, and replace it without having to get inside the coop.

Hinged Roofs and Walls: For smaller coops, consider installing a hinged roof or sidewalls that open up completely, giving you full access to the interior. This is especially useful for low-profile coops that do not have enough space to walk into.

Nesting Box Access

The nesting boxes are one of the areas that need frequent attention, especially if you are collecting eggs daily. Designing your coop with easy access to the nesting boxes will save you time and effort.

External Nesting Boxes: Consider building the nesting boxes on the outside of the coop with a hinged lid that allows you to collect eggs without entering the coop. This also makes cleaning the nesting boxes easier since you can open them from the outside and replace bedding quickly.

Removable Nesting Box Liners: Using removable liners in the nesting boxes, such as plastic trays or mats, makes cleaning much quicker. You can simply remove the liner, wash it off, and place it back inside the box.

The Deep Litter Method – A Low-Maintenance Approach

The deep litter method is one of the most popular strategies for reducing the amount of time spent cleaning a chicken coop. By allowing bedding material to accumulate and compost inside the coop, you can create a more self-sustaining system that benefits both you and your chickens.

How the Deep Litter Method Works

The deep litter method involves allowing bedding material, such as wood shavings, straw, or pine needles, to build up inside the coop over time. Rather than removing the bedding frequently, you add fresh material on top as needed. The bedding and chicken droppings will naturally compost, creating heat during the winter months and reducing the need for frequent cleaning.

Layering the Bedding: Start with a thick layer of dry, absorbent bedding on the coop floor. As the chickens drop manure, the bedding absorbs it, and you add more fresh bedding on top. The bottom layers begin to break down and compost naturally, providing warmth and reducing odors.

Turning the Bedding: Periodically, use a rake or pitchfork to turn the bedding and mix in the droppings. This encourages composting and keeps the bedding fresh longer.

Annual or Bi-Annual Cleanout: With the deep litter method, you only need to do a full cleanout once or twice a year. This makes it a much lower-maintenance option compared to traditional cleaning methods, where you might need to remove bedding weekly.

Benefits of the Deep Litter Method

The deep litter method offers several benefits, especially if you are looking to reduce the amount of time you spend cleaning the coop.

Reduced Cleaning Frequency: Instead of cleaning the coop every week, you can simply add fresh bedding as needed, reducing the overall maintenance.

Natural Heating: As the bedding and manure compost, they generate heat, which can help keep the coop warmer in the winter months, reducing the need for additional heating.

Free Fertilizer: After a few months, the composted bedding becomes a rich source of fertilizer that you can use in your garden. This adds to the sustainability of your setup.

Improved Chicken Health: The deep litter method creates a healthy microenvironment in the coop, with beneficial bacteria breaking down waste and reducing harmful pathogens.

Challenges of the Deep Litter Method

While the deep litter method is low-maintenance, it is not without its challenges.

Proper Ventilation: Since the bedding will be composting inside the coop, it is essential to have good ventilation to prevent moisture buildup and ammonia odors. Without adequate airflow, the coop can become too humid, which is harmful to your chickens.

Monitoring Moisture Levels: The bedding should remain dry and absorbent. If it becomes too wet, it will stop composting and may develop an unpleasant smell. you will need to monitor the bedding regularly to ensure it stays dry, especially in humid climates.

Waste Management and Drainage Solutions: One of the biggest challenges in maintaining a chicken coop is managing waste. Chickens produce a lot of droppings, and if they are not managed properly, they can lead to odor problems, moisture buildup, and even health issues for your flock.

Managing Manure in the Coop

Dealing with chicken manure can be one of the messiest parts of keeping chickens, but with the right strategies, you can minimize how often you need to clean and remove waste.

Roosting Bar Trays: Install trays or dropping boards under the roosting bars to catch manure overnight. Since chickens do most of their pooping while they are roosting, these trays will collect the majority of the droppings. You can simply scrape them off or dump the tray every few days, reducing the amount of waste that accumulates on the floor.

Composting Chicken Manure: Composting is one of the best ways to deal with chicken manure. Once removed from the coop, manure can be added to a compost pile along with kitchen scraps and yard waste. After a few months, the compost will break down into a nutrient-rich fertilizer that's perfect for the garden.

Drainage for Wet Areas

In addition to managing waste, it is important to ensure proper drainage in and around the coop to prevent water from pooling and creating unsanitary conditions. Wet bedding and standing water can lead to ammonia buildup, bad odors, and health problems for your chickens.

Sloped Floors and Runs: A sloped floor inside the coop and run helps direct water away from the structure, reducing the risk of puddles forming. Make sure the ground surrounding the coop is also sloped to prevent rainwater from running into the coop.

Gravel and Sand for Drainage: If you live in a rainy area or have problems with water pooling around the coop, consider adding a gravel or sand base to the floor of the run. These materials help with drainage and keep the ground dry. Sand is especially good for absorbing moisture and is easy to rake clean.

Essential Tools for Cleaning and Maintenance

Even with a well-designed coop, you will still need to do some regular cleaning and maintenance. Having the right tools on hand can make the job easier and faster.

Basic Cleaning Tools

To make cleaning as easy and efficient as possible, it is important to have a few key tools available. These will help you keep the coop clean without spending too much time or effort.

Shovel and Pitchfork: These tools are essential for removing old bedding, turning the deep litter, and managing waste. A pitchfork is particularly useful for handling straw or wood shavings, while a shovel works best for scraping up waste.

Scraper or Putty Knife: A scraper or putty knife is great for cleaning off dried droppings from roosts, walls, and nest boxes. It is an essential tool for keeping surfaces clean without having to do a deep scrub.

Broom and Dustpan: Keeping a broom and dustpan in the coop allows you to quickly sweep up any loose bedding or droppings. A push broom can be useful for larger coops with hard floors like concrete.

Disinfectants and Cleaning Solutions

In addition to removing waste and bedding, it is important to disinfect the coop regularly to prevent the buildup of harmful bacteria, mites, and other pests.

Vinegar and Water Solution: For a natural cleaning solution, a mixture of vinegar and water works well for disinfecting surfaces and neutralizing odors. It is non-toxic, safe for chickens, and effective against bacteria.

Mild Bleach Solution: In cases where you need to do a deep clean, a mild bleach solution (one part bleach to ten parts water) can help disinfect surfaces. Just be sure to rinse thoroughly and allow the coop to air out before letting the chickens back inside.

Designing for Drainage and Waste Reduction

A self-running coop should be designed not only to reduce cleaning but also to manage water and waste effectively. Proper drainage is key to keeping both the inside and outside of the coop clean and dry, while design features like sloped floors and raised nesting boxes can help reduce the amount of waste that accumulates.

Sloped Floors for Waste and Water Management

A sloped floor in the coop can help with both waste and water management. The slight incline will direct waste and spilled water toward a designated area, making it easier to clean up and reducing the risk of moisture buildup.

Drainage Channels: If your coop is in an area with heavy rainfall, consider installing drainage channels around the coop to direct water away from the structure. This will help prevent water from pooling inside and keep the bedding dry.

Raised Nesting Boxes and Roosts

Another way to reduce waste buildup is by raising the nesting boxes and roosting bars off the floor. This prevents chickens from soiling the boxes with droppings and makes cleaning underneath them easier.

Nesting Box Position: Place nesting boxes higher up in the coop, away from where chickens typically roost or perch. This keeps the bedding in the boxes cleaner and reduces the amount of time you will need to spend cleaning them out.

Designing your coop with cleaning and maintenance in mind can save you hours of work in the long run. By incorporating features like removable trays, good drainage, and the deep litter method, you can create a self-running coop that stays clean with minimal effort.

Designing the Run – Space and Shelter

The chicken run is where your flock will spend most of their time, so it is important to design it carefully to meet their needs for space, protection, and stimulation. A well-designed run will allow your chickens to forage, dust bathe, and exercise while keeping them safe from predators and the elements. When done right, a run can significantly reduce the time you spend on daily maintenance by preventing muddy, messy conditions and keeping your chickens healthier.

In this chapter, we will explore how to determine the right size for your run, the best materials to use for building, how to provide adequate shelter from the weather, and tips for predator-proofing your setup. By the end, you will be ready to create a spacious, safe, and easy-to-maintain chicken run that keeps both you and your flock happy.

Determining the Right Size for Your Run

One of the first considerations when designing a chicken run is determining how much space your chickens will need. Giving them plenty of room to roam not only improves their health and reduces stress, but it also keeps the run cleaner, as chickens will spread out instead of concentrating waste in one area.

Space Per Chicken

The recommended space for chickens in an outdoor run is about 8 to 10 square feet per bird. This gives them enough room to scratch, dust bathe, and forage without feeling overcrowded. Overcrowding can lead to stress, feather pecking, and increased disease transmission, so it is important to err on the side of more space whenever possible.

Minimum Space: For small backyard flocks, make sure each chicken has at least 8 square feet in the run. For example, a flock of 6 chickens should have a run that's at least 48 square feet (6x8 feet).

More Space, Less Maintenance: The more space you provide, the less frequently you will need to clean the run. A spacious run allows droppings to break down naturally and prevents the ground from becoming muddy or overrun with waste.

Vertical Space and Perches

In addition to horizontal space, chickens appreciate vertical space where they can perch. Adding perches or logs in the run allows your chickens to roost off the ground, which helps them feel more secure and prevents boredom.

Perch Placement: Install perches at different heights in the run, but make sure they are low enough that chickens can jump down without injury. Stagger the perches to prevent chickens from pooping on each other when roosting.

Natural Perches: You can use natural materials like tree branches or logs as perches, which not only provide enrichment but also help keep your chickens' feet healthy by giving them varied surfaces to stand on.

Materials for Building a Sturdy, Low-Maintenance Run

The materials you use to build your run will affect how much maintenance it requires and how well it protects your chickens from predators. You want to choose materials that are durable, weather-resistant, and predator-proof.

Fencing Materials – Predator Proofing and Durability

The fencing you choose for your run is perhaps the most important element in keeping your flock safe. While traditional chicken wire is often used for fencing, it is not the most secure option. Many predators, including raccoons, foxes, and dogs, can tear through chicken wire or squeeze through the gaps.

Hardware Cloth: The best material for predator-proof fencing is hardware cloth with a small mesh size (½-inch or less). Hardware cloth is much stronger than chicken wire and prevents even small predators, like weasels or rats, from getting into the run.

Pros: Strong, durable, and predator-proof. The small mesh size also keeps out smaller pests like snakes.

Cons: More expensive and harder to work with than chicken wire, but the investment pays off in long-term security.

Welded Wire Fencing: For large runs, welded wire fencing with larger gaps (around 2x4 inches) can be used in combination with hardware cloth along the bottom to provide security. This can save money on fencing material for larger areas while still offering good protection.

Pros: Less expensive than hardware cloth for covering large areas. Provides good protection against larger predators.

Cons: Smaller predators can slip through or dig under the fence, so it is best used alongside hardware cloth at ground level.

Flooring Materials for Easy Maintenance

Choosing the right ground cover for your run can reduce the amount of time you spend cleaning up and managing waste. Depending on your climate and how much time your chickens spend in the run, certain materials work better than others.

Gravel or Sand: Both gravel and sand are excellent options for the base of a chicken run. They promote good drainage, preventing mud from building up after rain, and are easy to rake or sift clean.

Gravel: Ideal for areas with poor drainage, gravel allows water to flow through and keeps the run from getting too muddy. It also deters predators from digging underneath the fence.

Sand: Sand is easy to clean and provides a natural dust bath for chickens, which helps with parasite control. Chickens enjoy scratching in sand, and it stays cooler in the summer.

Dirt: While dirt is the most natural option, it can turn into a muddy mess in wet weather, making the run harder to clean. If you use dirt, be sure to manage drainage effectively and add fresh material like straw or wood chips during rainy seasons.

Straw or Wood Chips: Straw or wood chips are commonly used in runs as they absorb moisture and provide a soft surface for chickens. However, they need to be changed regularly to prevent them from becoming wet and moldy, especially in rainy climates.

Straw: Works well in drier climates but can become a problem in wet conditions, where it may harbor mold and pests.

Wood Chips: More durable than straw, wood chips provide good drainage and can last longer before needing to be replaced.

Providing Shelter from Weather – Shade and Protection

Your chickens will need shelter in the run to protect them from harsh weather conditions. Whether it is the hot sun, pouring rain, or strong winds, providing adequate shelter will help keep your chickens healthy and comfortable.

Shade for Hot Weather

Chickens are highly susceptible to heat stress, so providing shade in the run is essential in hot climates or during summer months. A lack of shade can lead to overheating, stress, and even death in extreme cases.

Natural Shade: If possible, position your run near trees or other structures that provide natural shade. Trees not only block direct sunlight but also cool the air around the run.

Shade Cloth: If natural shade is not available, install shade cloth over part of the run to block out some of the sun's rays. Shade cloth is lightweight, easy to install, and does not trap heat, making it an excellent solution for hot climates.

Roof Extensions: Consider extending the roof of your coop over part of the run. This creates a shaded area that also protects chickens from rain and keeps the run drier.

Wind and Rain Protection

In addition to shade, your chickens will need protection from wind and rain. Wet chickens are more susceptible to illness, and a muddy run can become difficult to manage. Providing covered areas in the run helps your chickens stay dry and warm during bad weather.

Covered Areas: Adding a solid roof or partial covering to part of the run will give your chickens a place to escape from rain or snow. A corrugated metal or plastic roof works well for this and is easy to install over part of the run.

Windbreaks: In windy areas, you can create a windbreak using solid fencing or dense shrubs around part of the run. This will protect your chickens from strong gusts and help them stay warm during cold, windy days.

Predator-Proofing the Run

Predator-proofing your chicken run is one of the most important steps in keeping your flock safe. Chickens are vulnerable to a wide

range of predators, from neighborhood dogs to raccoons, hawks, and even weasels. A well-secured run will keep your chickens safe both day and night, reducing the risk of attacks.

Fencing and Burying Hardware Cloth

One of the best ways to protect your chickens from predators is to install hardware cloth around the run, especially at ground level where predators may try to dig under the fence.

Burying the Fence: Bury hardware cloth or wire mesh at least 12 inches below the ground around the perimeter of the run. This prevents predators from digging under the fence to get inside. You can also extend the mesh outward in an "L" shape to make digging even more difficult.

Apron Method: Instead of burying the fence, you can lay hardware cloth flat on the ground around the outside of the run. Secure it with stakes or rocks. Predators will try to dig but won't be able to get past the wire.

Covering the Run to Prevent Aerial Attacks

Hawks and other birds of prey pose a significant threat to chickens, especially in open runs. To protect your flock from aerial attacks, it is important to cover the top of the run.

Netting or Wire Roof: Install poultry netting, hardware cloth, or welded wire over the top of the run to prevent hawks and other birds of prey from swooping in. This also helps keep climbing predators, like raccoons or possums, from entering the run.

Roof Panels: For added protection and weatherproofing, consider covering part or all of the run with solid roofing panels, such as metal or plastic sheets. This keeps out both predators and rain, making the run safer and more comfortable for your chickens.

Enrichment and Entertainment for a Happy Flock

Keeping your chickens mentally and physically stimulated is key to preventing boredom, stress, and destructive behaviors like feather pecking. Incorporating enrichment features into the run will not only keep your flock happy but also make them healthier and more productive.

Dust Bathing Areas

Dust bathing is an important natural behavior for chickens. It helps them clean their feathers and control parasites like mites and lice. Providing a designated dust bathing area in the run will keep your chickens healthy and entertained.

Sand or Dirt Pit: Create a dust bathing area by filling a shallow pit or box with dry sand, dirt, or a mixture of both. Your chickens will regularly roll and flap in the dust to clean themselves and cool off.

Add Diatomaceous Earth: To further aid in parasite control, you can mix a small amount of diatomaceous earth into the dust bathing area. Diatomaceous earth is a natural substance that helps prevent mites and lice.

Foraging Areas and Feeders

Chickens love to forage, and providing opportunities for them to scratch and peck will keep them engaged and reduce boredom. You can also introduce enrichment feeders that encourage natural behaviors.

Rotating Foraging Areas: If possible, rotate the run to give your chickens access to fresh ground. This allows them to forage for insects, grass, and seeds, which not only keeps them entertained but also supplements their diet.

Hanging Treat Feeders: Hanging treat dispensers or suet feeders filled with fruits, vegetables, or grains can provide mental stimulation for your chickens. They will enjoy pecking at the hanging treats, keeping them occupied throughout the day.

Chicken Swings or Logs: Adding logs or even a simple chicken swing to the run gives your chickens something fun to perch on and explore. These small additions can make a big difference in keeping your flock entertained and content.

By designing a spacious, well-protected, and enriching run, you can create an outdoor space where your chickens thrive. A good run not only promotes better health and safety for your flock but also makes your daily chicken care easier and more enjoyable. With the right materials, layout, and predator-proofing strategies, your run will be a low-maintenance haven for your chickens.

Feeding Systems – Set It and Forget It

Feeding your chickens can be one of the more repetitive daily tasks, but it is also one of the easiest to automate. With the right feeding systems, you can ensure that your chickens have a consistent supply of food while reducing the time you spend on daily feeding routines. In this chapter, we'll explore different types of automated feeders, how to ensure a balanced diet for your flock, and tips for troubleshooting common feeding issues.

By implementing a low-maintenance feeding system, you will save time and ensure that your chickens are well-fed, even if you can't always be there to fill their feeders yourself.

Types of Automated Feeding Systems

There are several different types of automated feeding systems available, each with its own advantages and drawbacks. Whether you have a small backyard flock or a larger group of chickens, there's a feeding system that can work for your setup.

Gravity Feeders

Gravity feeders are one of the simplest and most popular options for automating chicken feeding. These feeders work by storing a large amount of feed in a hopper, which then dispenses the food into a tray as the chickens eat. As long as the hopper is full, the feeder will automatically keep the tray filled with food.

Pros: Gravity feeders are easy to use, relatively inexpensive, and require minimal setup. They can hold several days' worth of feed, depending on the size, which means you won't need to refill them every day.

Cons: Gravity feeders can sometimes become clogged if the feed is too moist or fine. Chickens may also scratch food out of the tray, leading to waste. Additionally, because they're open to the air, pests like rodents or wild birds may be attracted to the feeder.

Best for: Small to medium-sized flocks in backyard or enclosed setups where pest control is manageable.

Treadle Feeders

Treadle feeders are designed to dispense feed only when a chicken steps on a lever or platform, which opens the feeder's lid. This type of feeder helps reduce waste by keeping the feed covered when it is not being eaten and also deters pests like rats, mice, and wild birds.

Pros: Treadle feeders are great for preventing waste and keeping out pests. The covered design also keeps the feed dry and clean, which is especially helpful in wet or humid climates.

Cons: Some chickens may take a little time to learn how to use the treadle, especially if they are timid or young. Treadle feeders are also more expensive than gravity feeders and may require some training for the flock to use them effectively.

Best for: Flocks in areas where pests are a problem or for keepers looking to reduce feed waste. Ideal for outdoor or open coops.

Automatic Timed Feeders

Automatic timed feeders are more advanced systems that dispense feed at specific times during the day, usually controlled by a timer or sensor. These feeders are useful if you want to control feeding times, especially in cases where you want to limit the amount of feed available or avoid overfeeding.

Pros: Allows you to control feeding times and portions, which can help prevent overfeeding. They're useful if you need to feed your chickens at specific times but can't always be there to do it manually.

Cons: Timed feeders can be expensive and require a power source, either through electricity or batteries. They also need regular maintenance to ensure the mechanism does not get clogged or break down.

Best for: Larger flocks or those who want precise control over feeding times. They work well in environments where free feeding isn't practical.

DIY or Homemade Feeders

For those who prefer a hands-on approach, building a DIY feeder can be a cost-effective and customizable option. There are many simple designs that use basic materials like PVC pipes or buckets to create a gravity-fed system.

Pros: DIY feeders are inexpensive and customizable. You can build them to fit your space and flock size, and they're a great option if you are on a tight budget.

Cons: Homemade feeders may not be as durable or effective as commercial options, especially in terms of preventing pests or controlling feed waste. They may also require more maintenance and fine-tuning to work properly.

Best for: Chicken keepers who enjoy DIY projects and want a budget-friendly solution that can be tailored to their coop's layout.

Ensuring a Balanced Diet for Your Chickens

While automating the feeding process makes things easier for you, it is important to ensure that your chickens are receiving a balanced diet that meets all their nutritional needs. Chickens require a mix of protein, vitamins, minerals, and calcium to stay healthy and productive.

Types of Chicken Feed

Choosing the right type of feed is essential for keeping your chickens healthy, whether they're layers, meat birds, or ornamental breeds. Here's an overview of the main types of commercial chicken feed:

Starter Feed: This feed is designed for chicks and contains a higher level of protein (around 18-20%) to support rapid growth. It is typically fed until the chicks are about 8 weeks old.

Grower Feed: After the chicks transition from starter feed, they move on to grower feed, which contains slightly less protein (around 16-18%). This helps them continue growing without the risk of becoming overweight.

Layer Feed: Layer feed is formulated for hens that are actively laying eggs. It contains around 16% protein and added calcium to support strong eggshells. This is the main feed for adult laying hens and should be provided once they start laying, usually around 18 weeks of age.

All-Purpose or Flock Feed: For mixed flocks with different types of chickens, an all-purpose feed can be used. It provides balanced nutrition for both layers and non-layers but may need to be supplemented with additional calcium for laying hens.

Supplementing the Diet

In addition to their main feed, chickens often benefit from supplements like grit, calcium, and occasional treats.

Grit: Chickens do not have teeth, so they need grit to help grind up food in their gizzards. If your chickens are free-ranging, they'll likely find their own grit in the form of small stones and pebbles. If they're in a confined run, you will need to provide it.

Calcium: Laying hens need extra calcium to produce strong eggshells. You can provide this by offering crushed oyster shells or limestone grit in a separate feeder. Chickens will eat it as needed, ensuring they get the right amount of calcium.

Treats and Scraps: While treats should only make up a small portion of your chickens' diet, they enjoy kitchen scraps, greens, and grains. Just be careful not to overfeed treats, as this can throw off their nutritional balance.

Placement of Feeders for Maximum Efficiency

Where you place your feeders can impact not only how easy they are to maintain but also how your chickens interact with them. Strategic feeder placement can help reduce waste, prevent overcrowding, and ensure that all chickens have access to food.

Inside the Coop vs. Outside in the Run

Whether you place the feeder inside the coop or outside in the run depends on your setup and the weather conditions in your area.

Inside the Coop: Placing feeders inside the coop can protect the feed from rain and prevent wild birds or other pests from accessing it. It is a good option if your run isn't fully enclosed or predator-proof.

Outside in the Run: If you have a predator-proof run, placing the feeder outside gives chickens more space to move around and prevents the coop from becoming cluttered or dirty. It also encourages chickens to spend more time outside, reducing boredom and feather pecking.

Elevating Feeders to Reduce Waste

One of the simplest ways to reduce feed waste is to elevate the feeder off the ground. Chickens tend to scratch at their food, which can lead to a lot of wasted feed being kicked onto the ground. By raising the feeder to about the height of the chickens' backs, you can prevent them from spilling or scratching out feed.

Hanging Feeders: Hanging feeders from the roof or a sturdy post keeps them off the ground and reduces waste. This also keeps the feed cleaner by preventing dirt and bedding from getting kicked into the tray.

Platform Feeders: For larger or heavier feeders, consider placing them on a raised platform. This still elevates the feeder to reduce waste but provides more stability than hanging options.

Troubleshooting Common Feeding Problems

Even with an automated system, you may encounter some issues with feeding your chickens. Here are a few common problems and how to solve them.

Pests and Rodents

One of the biggest challenges with any feeding system is keeping pests like rats, mice, and wild birds from stealing food. Not only can pests eat large amounts of feed, but they also introduce diseases and parasites into your coop.

Use a Treadle Feeder: A treadle feeder is one of the best ways to keep pests out. Since the feed is only accessible when a chicken steps on the treadle, rats and mice are unable to reach it.

Store Feed Securely: Always store your feed in a secure, rodent-proof container with a tight-fitting lid. Metal bins are ideal for this, as they prevent rodents from chewing through the material.

Feed Waste

Feed waste can become expensive and messy if it is not managed properly. Chickens are notorious for scratching at their food, which can lead to a lot of wasted feed on the ground.

Elevate the Feeder: As mentioned earlier, raising the feeder off the ground helps reduce waste by preventing chickens from kicking feed out of the tray.

Limit Free Feeding: If you notice that your chickens are wasting a lot of feed, consider switching to a timed feeder or limiting the amount of food available at any given time. This can prevent overeating and reduce the amount of feed that gets spilled.

Uneven Access to Feed

In larger flocks, dominant chickens may bully others away from the feeder, leading to uneven access to food. This can result in some chickens not getting enough to eat while others overeat.

Provide Multiple Feeders: If you have a larger flock or notice bullying around the feeder, adding a second feeder can help ensure that all chickens have access to food. Place the feeders in different areas of the coop or run to prevent competition.

Monitoring Feed Consumption

Even with an automated system, it is important to regularly monitor how much feed your chickens are consuming. Sudden changes in consumption can be a sign of health issues or problems with the feeder itself.

Track Feed Usage: Keep an eye on how much feed your chickens are eating each day. If you notice a significant drop in consumption, it could indicate illness, stress, or a problem with the feeder.

Check for Clogs: Occasionally, feeders—especially gravity or timed feeders—can become clogged with moist or powdery feed. Make sure to check your feeders regularly to ensure they're dispensing food properly.

By setting up a reliable, low-maintenance feeding system, you will free up your time while ensuring your chickens are well-fed and healthy. Whether you choose a gravity feeder, treadle feeder, or more advanced timed system, automating this essential task will keep your flock happy and reduce your daily workload.

Watering Solutions for the Busy Keeper

Water is just as important as food for your chickens, and ensuring they have a constant supply of clean, fresh water is essential for their health and productivity. However, keeping waterers full and clean can be a time-consuming task, especially if you have a large flock. Automating the watering system for your coop can save you hours of work and ensure your chickens never run out of water, even on the hottest days.

In this chapter, we will explore various types of automated watering systems, how to keep water fresh and clean, solutions for different climates, and troubleshooting common watering issues. By the end, you will have all the tools you need to set up a reliable, low-maintenance watering system that works year-round.

Types of Automated Watering Systems

There are several different types of automated watering systems available, ranging from simple gravity-fed designs to more advanced systems that use automatic refills. Choosing the right system depends on the size of your flock, your climate, and how hands-off you want the system to be.

Nipple Waterers

Nipple waterers are one of the most popular options for automated chicken watering systems. These systems use small metal or plastic nipples that release water when the chickens peck at them. The water is stored in a reservoir (such as a bucket or tank) and dispensed only when needed, keeping it clean and reducing waste.

Pros: Nipple waterers are highly sanitary because the water stays inside the reservoir, only coming out when the chickens activate the nipples. This reduces the risk of contamination from dirt, bedding, or droppings. They also conserve water, as very little is wasted or spilled.

Cons: Chickens need to learn how to use the nipple system, which can take a bit of time. In cold climates, nipple waterers can freeze, so you will need to find a way to keep the water warm.

Best for: Medium to large flocks or chicken keepers who want a clean, low-maintenance system. Works best in climates where freezing temperatures aren't an issue, or where heating can be added.

Cup Waterers

Cup waterers operate similarly to nipple waterers but use a small cup that fills with water when chickens peck at a valve. As with nipple

systems, the water is stored in a reservoir and only dispensed when needed, keeping it fresh and clean.

Pros: Cup waterers are easy for chickens to learn how to use, and they allow for small amounts of water to be available at a time, reducing waste and spills. They're also easy to clean and maintain.

Cons: Like nipple waterers, cup waterers can freeze in cold climates. They also need to be checked regularly to ensure the cups do not become clogged with dirt or bedding.

Best for: Smaller flocks or those in climates where freezing isn't a major issue. Ideal for chicken keepers who prefer a more hands-on system that's still automated.

Gravity-Fed Waterers

Gravity-fed waterers are one of the simplest types of automated systems. These waterers work by holding water in an elevated container, which dispenses it into a trough or basin as the chickens drink. As the water level in the basin lowers, gravity pulls more water from the reservoir to refill it.

Pros: Gravity-fed waterers are easy to set up, inexpensive, and require very little maintenance. They can be built using common materials like buckets, PVC pipe, or troughs, making them a great DIY project.

Cons: These systems can be prone to contamination if chickens kick dirt or bedding into the water basin. In freezing temperatures, the water can freeze and block the flow.

Best for: Small to medium-sized flocks in moderate climates where freezing and contamination aren't major concerns.

Automatic Waterers with a Float Valve

Automatic waterers connected to a float valve system refill automatically as the water level drops, making them a fully hands-off option. These systems are typically connected to a water supply, such as a hose or a rainwater collection system, and the float valve regulates the water level in a trough or basin.

Pros: These systems are ideal for chicken keepers who want a fully automated watering system. you will never need to refill water manually, and as long as the water supply is uninterrupted, your chickens will always have access to fresh water.

Cons: Float valve systems require a continuous water supply, which may not be practical in all locations. If the system malfunctions or becomes clogged, the waterer can overflow or stop filling entirely. In cold climates, the water supply can freeze, rendering the system useless without a heating solution.

Best for: Large flocks or farms with access to a continuous water supply. Ideal for those who want minimal involvement in the watering process.

Keeping Water Fresh and Clean

No matter what type of watering system you use, keeping the water clean and fresh is essential for your chickens' health. Contaminated or dirty water can lead to disease, reduced egg production, and dehydration if your chickens refuse to drink it.

Protecting Water from Contamination

One of the biggest challenges in keeping water clean is preventing contamination from dirt, droppings, and bedding. Automated systems like nipple and cup waterers help prevent contamination, but if you are using a gravity-fed or open water system, you will need to take additional steps.

Elevate the Waterer: Placing the waterer at the height of the chickens' backs will prevent them from scratching dirt or bedding into the water. It also discourages them from stepping in it or knocking it over.

Covered Water Basins: If you are using an open water basin or trough, consider adding a cover or partial lid to reduce the amount of dirt and debris that falls into the water.

Regular Cleaning: Even automated systems need occasional cleaning to prevent algae or bacteria buildup. Make it a habit to check and clean your waterers weekly, especially in warmer weather when bacteria can multiply quickly.

Preventing Algae Growth

In hot, sunny climates, algae growth can be a problem, particularly in open or transparent waterers. Algae not only make the water unappealing to chickens but can also harbor harmful bacteria.

Shade the Waterer: Placing the waterer in a shaded area or using opaque containers can help reduce algae growth. Algae thrive in sunlight, so keeping the waterer out of direct sunlight can slow its development.

Use Apple Cider Vinegar: Adding a small amount of apple cider vinegar to the water can help prevent algae growth. About a tablespoon of vinegar per gallon of water is sufficient to discourage algae without affecting the taste of the water for the chickens.

Regular Cleaning: Scrubbing the inside of the waterer with a brush every week or two helps prevent algae from building up. Make sure to use a non-toxic cleaner or just plain water to avoid contaminating the water supply.

Cold Weather Considerations – Preventing Frozen Water

One of the biggest challenges for chicken keepers in colder climates is preventing water from freezing during the winter. Frozen waterers can quickly leave your flock without access to water, which is not only inconvenient but dangerous for your chickens.

Heated Waterers

The easiest way to prevent freezing is to invest in a heated waterer. Heated waterers use electricity to keep the water at a temperature above freezing, ensuring your chickens always have access to liquid water, even in subzero temperatures.

Pros: Heated waterers take the hassle out of winter chicken care by ensuring the water stays thawed without needing constant attention.

Cons: Heated waterers require a power source, so they may not be practical for all setups. They also add an additional cost to your monthly electricity bill.

Best for: Chicken keepers in areas with prolonged freezing temperatures who want a hands-off solution for winter watering.

Heated Bases for Existing Waterers

If you already have a waterer that you like, you can purchase a heated base that sits under the waterer and keeps it from freezing. These heated bases are a great option if you do not want to replace your current system but still need a way to prevent freezing.

Pros: Heated bases are versatile and can be used with most types of waterers. They provide a simple solution without the need to buy a completely new system.

Cons: Like heated waterers, heated bases require a power source, and they only prevent freezing—they do not keep the water warm. If the temperature drops too low, the water may still freeze around the edges.

Best for: Those who want to use their existing waterers in winter without having to worry about freezing.

Solar-Powered Solutions

For chicken keepers without access to electricity, solar-powered heated waterers are a great option. These systems use solar panels to generate enough heat to keep the water from freezing during the day, although they may struggle during prolonged periods of darkness or heavy snow.

Pros: Solar-powered waterers are ideal for off-grid setups and do not require access to electricity. They're also environmentally friendly.

Cons: Solar-powered systems may not be reliable in areas with limited sunlight during the winter months. Heavy snow or cloudy days can reduce their effectiveness, and they may not work as well overnight when temperatures drop the most.

Best for: Off-grid homesteads or chicken keepers in sunny, cold climates who want an eco-friendly solution for winter watering.

Rainwater Collection Systems

One of the most sustainable ways to provide water for your chickens is by using a rainwater collection system. These systems capture rainwater from the roof of your coop or another structure and store it in a tank or barrel, which can then be connected to your waterers.

Setting Up a Rainwater Collection System

A basic rainwater collection system consists of a gutter, downspout, and a collection barrel. The rainwater is collected from the roof and funneled into the barrel, where it can be stored until needed.

Gutter Installation: Install gutters along the edge of the coop's roof to catch rainwater. The gutter should lead to a downspout that directs water into the collection barrel.

Collection Barrel: Use a food-grade barrel or tank to collect and store the rainwater. It is important to cover the barrel to keep out debris and mosquitoes.

Water Purification: Rainwater is usually safe for chickens to drink, but it can be contaminated by dirt or leaves from the roof. Installing a simple filter or adding a first-flush diverter, which allows the first few gallons of rain to wash away dirt before the water enters the barrel, can help keep the water clean.

Connecting Rainwater to Watering Systems

Once you have collected the rainwater, you can connect it to an automated watering system like a nipple or cup waterer. Simply attach a hose or pipe to the bottom of the collection barrel and connect it to your waterers.

Float Valve Connection: If you are using an automatic float valve system, you can connect the rainwater barrel directly to the float valve. The system will automatically refill the waterer as the chickens drink, ensuring a constant supply of fresh water.

Gravity-Feed Connection: If your waterers are gravity-fed, you can position the rainwater barrel above the waterers and use gravity to direct water into the system as needed.

Troubleshooting Common Watering Problems

Even with an automated system, you may run into some common watering issues. Here are a few problems you might encounter and how to solve them.

Clogged Waterers

Over time, nipple or cup waterers can become clogged with dirt or debris, especially if the water source isn't perfectly clean. Clogged waterers can stop the flow of water and leave your chickens without access to water.

Regular Cleaning: Make sure to clean the waterers regularly to prevent clogs. A small brush or toothpick can be used to clear out any blockages in the nipples or cups.

Check the Water Source: If you are using rainwater or another non-filtered source, install a filter or screen to keep debris from entering the system and clogging the waterers.

Freezing Water

Even with heated waterers, freezing can still be an issue during very cold weather. If your waterers freeze, your chickens will be left without access to water, which can lead to dehydration.

Increase Heating: If your heated waterer or base isn't preventing freezing, you may need to increase the heat output or switch to a more powerful system.

Use Insulation: Wrapping the waterer or its reservoir in insulating material can help keep the water warmer for longer, especially overnight when temperatures drop.

By setting up an efficient watering system, you will ensure your chickens always have access to clean, fresh water with minimal effort on your part. Whether you choose a simple gravity-fed waterer or a more advanced automatic system, you will save time and keep your flock hydrated and healthy, no matter the season.

Lighting and Egg Collection Automation

Proper lighting plays a crucial role in maintaining egg production, especially during the shorter days of winter when natural daylight is limited. By automating the lighting inside your chicken coop, you can ensure consistent egg production throughout the year. Additionally, automating the egg collection process can save you time and reduce the risk of broken or dirty eggs.

In this chapter, we will explore the importance of lighting for egg production, the types of automated lighting systems available, and how to install a simple but effective egg collection system. Whether you are raising chickens for personal use or selling eggs, thcsc systems will help you maximize efficiency and minimize effort.

The Importance of Lighting for Egg Production

The laying cycles of chickens are heavily influenced by the amount of daylight they receive. To maintain healthy egg production, hens need about 14 to 16 hours of light each day. During the summer months, this is usually not an issue, but as the days shorten in winter, egg production often declines. By adding supplemental lighting to your coop, you can maintain consistent light levels and keep your hens laying year-round.

How Light Affects Laying Cycles

The reproductive systems of chickens are regulated by light exposure, specifically the length of daylight hours. When hens receive enough light, their bodies produce the hormones necessary for egg production. However, when daylight hours drop below 12-14 hours, their bodies begin to slow or stop egg production as a natural response to the shorter days of winter.

Impact on Productivity: If your goal is to maintain consistent egg production, especially in colder months, lighting is essential. Without it, many hens will reduce or stop laying until the days naturally lengthen again.

Molt and Egg Laying: In addition to shorter days, chickens also go through a natural molting period in which they lose and regrow their feathers. This often coincides with a drop in egg production. While adding light can help boost production, it is important to allow your hens to go through their molt without forcing them to continue laying, as this is a natural part of their cycle.

Light Requirements for Chickens

To keep egg production high, hens need about 14 to 16 hours of light daily. During winter, when natural daylight may only last 8-10 hours, supplemental lighting can be added to make up the difference.

Light Intensity: Chickens do not need overly bright light—about the same intensity as a 40-watt bulb is sufficient. You do not need industrial-strength lighting to maintain production.

Timing of Light.:It is important to gradually add light in the morning or evening rather than leaving the lights on all day. Abrupt changes in light levels can confuse the chickens and disrupt their natural cycles.

Automated Lighting Systems

Automating the lighting in your chicken coop can save you the hassle of manually turning lights on and off each day. Automated lighting systems ensure that your chickens receive consistent lighting without you having to remember to flip a switch.

Timers for Automatic Lighting

The simplest way to automate your coop's lighting is to use a timer. A timer can be set to turn the lights on before sunrise or stay on for a few hours after sunset, ensuring your chickens get the extra light they need without you having to monitor it constantly.

Pros: Timers are inexpensive, easy to install, and highly reliable. You can set them and forget them, making them a great low-maintenance solution.

Cons: Timers can not adjust for changes in daylight hours throughout the year. You may need to adjust the timer occasionally as the days get shorter or longer.

Best for: Chicken keepers who want a simple, affordable solution for maintaining consistent lighting without daily manual adjustments.

Photocell Sensors for Daylight Control

Photocell sensors offer a more sophisticated solution for automated lighting. These sensors detect the amount of natural daylight and turn the lights on or off based on the available light. For example, the lights can turn on automatically as the sun sets and turn off when natural daylight returns in the morning.

Pros: Photocell sensors automatically adjust to seasonal changes in daylight, making them ideal for year-round use without needing manual adjustment.

Cons: Photocell systems are more expensive than simple timers, and they require a bit more setup. Additionally, in some climates, low-light winter days may not provide enough natural light, leading to the lights staying on longer than necessary.

Best for: Chicken keepers who want a fully automated system that adjusts to changing light conditions without needing regular adjustments.

Solar-Powered Lighting Solutions

For off-grid setups or chicken coops without easy access to electricity, solar-powered lights are an excellent option. These systems use solar panels to collect energy during the day and power LED lights in the coop at night. Many solar-powered systems come with built-in timers or sensors for easy automation.

Pros: Solar-powered lights are eco-friendly, energy-efficient, and require no external power source. They are ideal for coops in remote areas or for those looking to reduce their energy costs.

Cons: Solar-powered systems may not generate enough energy during cloudy or short winter days, leading to inconsistent lighting. You may need a backup power source in some climates.

Best for: Off-grid homesteads or chicken keepers who want to reduce energy usage and do not have reliable access to electricity.

Egg Collection Automation

While automating your lighting ensures consistent egg production, automating the egg collection process can save you time and effort. Collecting eggs daily is essential to prevent breakage, keep the eggs clean, and maintain optimal freshness, but bending down to gather eggs from multiple nesting boxes can quickly become a chore.

Rollaway Nesting Boxes

Rollaway nesting boxes are designed to collect eggs automatically after they're laid, reducing the risk of breakage and keeping the eggs clean. In these boxes, the eggs gently roll down a slight incline into a collection area, where they stay safe and easy to gather.

Pros: Rollaway boxes prevent eggs from being broken, dirtied, or eaten by chickens. They also simplify egg collection by consolidating the eggs in one easy-to-access area.

Cons: Rollaway boxes are more expensive than standard nesting boxes and may require some adjustment to ensure the eggs roll smoothly without damage.

Best for: Chicken keepers who want to streamline egg collection, especially if they have a larger flock or prefer not to check nesting boxes multiple times a day.

DIY Egg Collection Systems

For those who enjoy a hands-on approach, DIY egg collection systems can be a cost-effective and customizable option. You can create a simple rollaway nesting box using materials like plywood, wire mesh, and foam padding to cushion the eggs as they roll into the collection tray.

Pros: DIY systems allow you to customize the size, shape, and design of the nesting boxes to fit your coop. They can be built on a budget and modified as needed.

Cons: Building a DIY egg collection system requires time and some basic carpentry skills. There's also the potential for trial and error as you adjust the incline and padding to ensure the eggs roll smoothly without cracking.

Best for: Chicken keepers who enjoy DIY projects and want a low-cost solution for automating egg collection.

Maintaining Clean and Sanitary Eggs

Even with an automated egg collection system, it is important to ensure that your eggs stay clean and sanitary. Dirty eggs can harbor bacteria, which can contaminate the eggshell and reduce the quality of your eggs.

Providing Clean Bedding

One of the easiest ways to keep eggs clean is by providing fresh, clean bedding in the nesting boxes. Straw, wood shavings, or nesting pads should be replaced regularly to prevent droppings or dirt from soiling the eggs.

Nesting Box Liners: Using washable nesting box liners or nesting pads can make it easier to keep the boxes clean. These liners can be removed, cleaned, and replaced regularly, ensuring the nesting area stays sanitary.

Changing Bedding: In addition to liners, make sure to change out bedding at least once a week, or more often if you notice it becoming soiled. Clean, dry bedding helps reduce the risk of dirty or broken eggs.

Egg Collection Frequency

Even with an automated egg collection system, it is important to check and collect eggs daily. Collecting eggs frequently reduces the chances of eggs being broken, eaten by chickens, or soiled by droppings.

Morning and Evening Collection: For the freshest eggs, try to collect them twice a day—once in the morning and once in the evening. This ensures that any eggs laid during the day are gathered before chickens can step on or soil them.

Using Rollaway Systems: If you are using a rollaway nesting box, you may only need to collect the eggs once a day, as the eggs will stay protected in the collection tray until you are ready to gather them.

Troubleshooting Common Lighting and Egg Collection Problems

While lighting and egg collection automation can greatly reduce your workload, occasional issues may arise. Here's how to troubleshoot some common problems.

Chickens Not Using Nesting Boxes

If you notice that your chickens are laying eggs outside the nesting boxes or on the floor, there may be an issue with the design or placement of the boxes.

Nesting Box Placement: Make sure your nesting boxes are in a quiet, dimly lit area of the coop. Chickens prefer to lay in secluded areas where they feel safe and comfortable.

Sufficient Boxes: Ensure you have enough nesting boxes for your flock. A general rule of thumb is one nesting box for every 4-5 hens. Overcrowded boxes can lead to chickens laying in undesirable locations.

Clean and Comfortable Bedding: Chickens may avoid nesting boxes if the bedding is dirty or uncomfortable. Keep the bedding fresh and soft to encourage hens to use the boxes.

Lighting Malfunctions

If your automated lighting system is not working correctly, you may need to troubleshoot the timer or sensor.

Check the Timer: If the lights are not turning on or off at the correct times, double-check the timer settings to ensure they're properly programmed. It is also a good idea to replace the batteries in battery-operated timers regularly.

Inspect the Photocell Sensor: If you are using a photocell sensor, ensure that the sensor is positioned where it can accurately detect daylight. Dirt or debris on the sensor can cause it to malfunction, so clean it regularly.

By automating your lighting and egg collection systems, you will create a more efficient coop that maximizes egg production while minimizing daily chores. Whether you opt for a simple timer for your lights or invest in a rollaway egg collection system, these automated solutions will help you save time and ensure your chickens are happy and productive year-round.

Building a Healthier Environment – Ventilation and Airflow

One of the most critical but often overlooked aspects of a well-designed chicken coop is proper ventilation. Good airflow keeps the air inside the coop fresh, reduces the buildup of moisture and ammonia, and prevents respiratory issues in your flock. Without adequate ventilation, your chickens may face health problems, and the coop can become a breeding ground for bacteria and mold.

In this chapter, we will explore the importance of ventilation, how to design a coop that promotes good airflow, and tips for managing humidity and odors year-round. A well-ventilated coop ensures a healthier environment for your chickens, reducing your maintenance workload while keeping your flock happy and productive.

Why Ventilation Is Essential in a Chicken Coop

Ventilation is crucial for maintaining air quality inside the coop. Chickens produce moisture, ammonia, and heat through their breathing and droppings, and without proper airflow, these can build up to harmful levels. A well-ventilated coop allows fresh air to circulate while expelling moisture, odors, and heat, creating a healthier living environment for your chickens.

Preventing Respiratory Issues

Chickens are prone to respiratory problems, and poor ventilation is one of the primary causes. When moisture builds up inside the coop, it can lead to damp conditions that encourage the growth of mold, bacteria, and harmful ammonia gases from chicken droppings.

Ammonia Buildup: Ammonia from chicken manure can irritate chickens' respiratory systems, leading to infections, eye irritation, and reduced egg production. Proper ventilation helps dilute ammonia and prevent it from reaching dangerous levels.

Dust and Dander: Chickens naturally produce dust from their feathers and bedding, which can accumulate in a poorly ventilated coop. Dusty conditions can irritate their lungs, causing respiratory issues. Good airflow helps reduce dust levels inside the coop.

Controlling Humidity and Moisture

Chickens produce moisture through their breath and droppings, and without proper ventilation, that moisture can lead to damp conditions. Dampness inside the coop increases the risk of frostbite in cold weather and promotes mold and bacteria growth, which can harm your chickens' health.

Humidity in Hot Climates: In warm climates, excess moisture can make the coop feel muggy, which can lead to heat stress and a higher risk of bacterial infections. Ventilation helps remove humidity and keeps the air inside the coop cooler and drier.

Moisture in Cold Climates: In colder climates, too much moisture can increase the risk of frostbite, especially on the combs and wattles. Ventilation reduces moisture levels, preventing frostbite and keeping the coop dry in winter.

Reducing Odors and Bacteria

Proper ventilation helps manage odors by preventing the buildup of ammonia and stale air inside the coop. A well-ventilated coop will smell fresher, which benefits both your chickens and anyone nearby. Additionally, better airflow reduces the risk of bacteria and mold growing inside the coop, which can be harmful to both chickens and humans.

Designing for Optimal Airflow

When designing a chicken coop, it is important to plan for airflow from the start. There are several design features that can help promote good ventilation, including windows, vents, and doors. The goal is to allow fresh air to flow through the coop while expelling heat, moisture, and odors.

Natural Ventilation – Windows and Vents

Natural ventilation relies on the natural movement of air through windows and vents. By strategically placing openings in your coop, you can create a system where fresh air enters, and stale air escapes without the need for fans or other mechanical systems.

Windows: Windows are one of the easiest ways to add ventilation to your coop. They allow natural light to enter while providing airflow. In hot climates, windows should be placed on opposite sides of the coop to create cross-ventilation, allowing air to flow through and cool the coop.

Wire Mesh Screens: All windows should be covered with hardware cloth or wire mesh to keep predators out while allowing air to flow freely.
Vents: Vents, particularly placed high on the walls or at the roofline, help expel hot air and moisture. Since hot air rises, vents located near the top of the coop are essential for releasing heat in the summer and reducing moisture in the winter.

Roof Ridge Vents: Installing a ridge vent along the peak of the roof is an effective way to allow hot, humid air to escape. Ridge vents are particularly useful in climates with hot summers, as they help keep the coop cooler.

Gable Vents: Gable vents are triangular vents located near the roofline on the side walls of the coop. They allow heat and moisture to escape while drawing in cooler, fresh air from lower openings.

Cross-Ventilation for Year-Round Fresh Air

Cross-ventilation is the movement of air through the coop from one side to the other. To achieve this, you will need to place windows, vents, or doors on opposite sides of the coop. In hot weather, cross-ventilation is essential for cooling the coop, while in cold weather, it helps keep moisture levels down without creating drafts near the chickens.

Summer Ventilation: In the summer, open windows and vents should allow air to flow freely through the coop, keeping it cool. Just be sure to provide enough shade and avoid drafts directly on your chickens, especially at night when temperatures can drop.

Winter Ventilation: Even in winter, ventilation is necessary to prevent moisture buildup, but the goal is to avoid direct drafts on the chickens. Vents placed high on the walls or roof can remove moisture and ammonia without chilling your chickens. Close windows or low-level vents during the winter to prevent cold drafts.

Managing Humidity and Temperature

While ventilation helps control humidity and temperature, there are additional strategies to ensure your coop remains comfortable year-round. In both hot and cold climates, proper insulation and the strategic use of fans or heaters can help regulate conditions inside the coop.

Insulating the Coop

Insulation helps maintain a stable temperature inside the coop, keeping it warmer in the winter and cooler in the summer. Insulation can also reduce moisture buildup by preventing condensation on the walls and roof.

Materials for Insulation: Common insulation materials for coops include foam boards, fiberglass, and spray foam. For a more natural approach, you can use straw bales or wood shavings as insulating materials.

Where to Insulate: Insulate the walls, ceiling, and floor of the coop for maximum benefit. Focus on the areas that are most exposed to extreme temperatures, such as the roof and north-facing walls.

Using Fans for Extra Airflow

In particularly hot climates, natural ventilation may not be enough to keep the coop cool. Adding fans can help circulate air and prevent heat stress in your flock.

Solar-Powered Fans: Solar-powered fans are a great option for chicken keepers without access to electricity. They help move air through the coop without adding to your energy costs. Place the fan near a vent or window to help expel hot air and bring in cooler air from outside.

Oscillating Fans: In larger coops, oscillating fans can be used to circulate air. These fans should be placed in a location where they move air without blowing directly on the chickens, as drafts can cause stress.

Dehumidifiers for Humid Climates

In very humid climates, excess moisture can still be an issue even with good ventilation. A small dehumidifier can help remove moisture from the air, keeping the coop drier and reducing the risk of mold or bacterial growth.

Portable Dehumidifiers: These are useful in small or medium-sized coops where humidity is a persistent problem. Place the dehumidifier near a vent to help expel the moisture it collects.

Reducing Odors and Ammonia Buildup

Odor control is a major concern for chicken keepers, particularly in hot or humid climates where ammonia from chicken droppings can quickly build up. A combination of good ventilation, regular cleaning, and strategic bedding choices can help reduce unpleasant odors in your coop.

Managing Ammonia Levels

Ammonia is produced by the breakdown of chicken manure, and if it builds up inside the coop, it can harm your chickens' respiratory systems. Ventilation helps reduce ammonia levels, but there are additional strategies you can use to keep them in check.

Deep Litter Method: Using the deep litter method in your coop can help manage ammonia buildup. The bedding material, such as wood shavings or straw, absorbs moisture and helps break down manure, reducing the amount of ammonia released into the air.

Bedding Choices: Use absorbent bedding materials like wood shavings, straw, or hemp to help control moisture and reduce ammonia odors. Changing the bedding regularly also keeps the coop smelling fresher.

Regular Cleaning Routines

While good ventilation helps manage odors, regular cleaning is still necessary to keep the coop smelling fresh and reduce ammonia buildup.

Cleaning Frequency: Clean out the coop every week or two, depending on the size of your flock and the amount of manure produced. Remove wet or soiled bedding, especially around the roosting bars and nesting boxes.

Disinfecting the Coop: Periodically disinfect the coop to reduce the buildup of bacteria and prevent odors from becoming overwhelming. Use a mild vinegar solution or a non-toxic disinfectant designed for poultry coops.

Troubleshooting Common Ventilation Problems

Even with a well-designed ventilation system, you may encounter some common issues related to airflow and air quality. Here's how to troubleshoot them.

Poor Air Circulation

If your coop feels stuffy or you notice a buildup of moisture, your ventilation system may not be circulating air effectively.

Solution: Check the placement of your vents and windows. You may need to add more vents, especially at the roofline, to improve air circulation. Make sure that nothing is blocking the vents or windows, such as roosting bars, nesting boxes, or other coop features.

Cold Drafts in Winter

If you notice cold drafts near your chickens during the winter, it may mean that your ventilation system is allowing too much cold air to enter the coop.

Solution: Close any low-level vents or windows during the winter months. Focus on high vents near the roofline to expel moisture without causing drafts near the chickens. You can also use insulated covers for windows to reduce drafts while still allowing airflow.

Excessive Humidity or Moisture

If your coop is consistently damp or humid, even with ventilation, the issue may be related to excess moisture in the environment or poor drainage around the coop.

Solution: Check for leaks in the roof or walls that could be letting water into the coop. Improve drainage around the coop by sloping the ground away from the structure, and consider adding gravel or sand to the floor of the run to reduce mud and moisture.
By designing your coop with good ventilation and airflow, you will create a healthier, more comfortable environment for your chickens. Proper ventilation helps reduce the risk of respiratory issues, controls humidity, and keeps odors under control, making your coop a more pleasant place for both your flock and yourself.

Pest and Predator Management for a Self-Sustaining Coop

One of the biggest challenges of raising chickens is protecting them from pests and predators. Whether it is tiny mites infesting the coop or larger predators like raccoons and foxes, keeping your flock safe is essential to a successful, self-sustaining chicken operation. With the right design, prevention strategies, and management techniques, you can minimize the risks and keep your chickens healthy and secure with minimal intervention.

In this chapter, we'll explore common pests and predators that threaten chickens, as well as practical, low-maintenance solutions to keep them at bay. From physical barriers to natural deterrents, you will learn how to safeguard your coop and maintain a safe environment for your flock.

Common Chicken Pests and How to Manage Them

There are several types of pests that can cause problems for chickens, from mites and lice to rodents. Understanding the signs of an infestation and knowing how to prevent and treat these pests can make a big difference in the health of your flock.

Mites and Lice

Mites and lice are tiny parasites that live on chickens, feeding on their blood, skin, or feathers. They can cause discomfort, stress, and even illness if left untreated. Mites tend to hide in cracks and crevices in the coop during the day, coming out at night to feed on the chickens.

Symptoms: Chickens infested with mites or lice may scratch excessively, lose feathers, develop pale combs and wattles, or appear lethargic. You might also see mites or lice crawling on their skin or feathers, especially around the vent area.

Prevention: Keeping the coop clean and dry is key to preventing mite and lice infestations. Regularly changing bedding, using diatomaceous earth in the bedding, and providing dust bathing areas can help control these parasites.

Diatomaceous Earth (DE): Food-grade DE is a natural, non-toxic powder made from fossilized remains of diatoms. It helps dehydrate mites and lice when sprinkled in bedding or dust bathing areas. Chickens will dust-bathe in it, helping to control parasites on their bodies.

Treatment: If you notice an infestation, treat your chickens and coop immediately. Commercial poultry dust or sprays containing permethrin can be used to kill mites and lice. Be sure to treat both the chickens and the coop, focusing on cracks and crevices where mites may hide.

Flies and Mosquitoes

Flies and mosquitoes can be a nuisance in the coop, especially in warm or wet climates. They not only annoy chickens but can also spread diseases, such as avian pox, through their bites.

Prevention: Keep the coop clean and dry to reduce the presence of flies. Removing wet bedding, using fly traps, and installing screens over windows and vents can help. Stagnant water sources should be eliminated to prevent mosquitoes from breeding.

Natural Repellents: Planting herbs like mint, lavender, and basil around the coop can help deter flies and mosquitoes naturally. Hanging flypaper or using natural fly traps can also reduce the fly population.

Rodents (Rats and Mice)

Rats and mice are drawn to chicken coops in search of food and shelter. Once they invade, they can contaminate feed, chew through wiring, and spread diseases. Additionally, they can attract larger predators to the area.

Prevention: The best way to prevent rodents is by making sure the coop and run are rodent-proof. Store feed in metal containers with tight lids, and clean up any spilled feed promptly. Regularly check for gaps or holes in the coop walls or foundation and seal them with hardware cloth or other materials.

Rodent-Proof Feeders: Using treadle feeders, which require chickens to step on a platform to access feed, can help keep rodents from stealing food. Hanging feeders off the ground can also reduce access for rodents.

Traps and Bait: If you already have a rodent problem, setting traps around the outside of the coop can help. Avoid using poison, as it can harm your chickens or other wildlife. Instead, opt for snap traps or live traps placed in areas where chickens can't access them.

Common Chicken Predators and How to Protect Against Them

In addition to pests, your chickens are vulnerable to a wide range of predators. Raccoons, foxes, hawks, and even neighborhood dogs can pose a threat. Protecting your chickens with physical barriers and deterrents is essential to keeping your flock safe.

Raccoons, Foxes, and Coyotes

Raccoons, foxes, and coyotes are among the most common chicken predators. These animals are skilled at finding weak points in coop design, and they're known for digging, climbing, and squeezing through small gaps to get to chickens.

Predator-Proof Fencing: To prevent digging, bury hardware cloth or welded wire fencing at least 12 inches underground around the perimeter of the coop and run. This makes it difficult for predators to dig their way in. Another option is to create a "predator apron" by laying the hardware cloth flat on the ground around the run and securing it with stakes.

Hardware Cloth vs. Chicken Wire: Chicken wire is not predator-proof; many predators can tear through it easily. Always use

hardware cloth or welded wire for fencing to provide a more secure barrier.
Securing the Coop: Make sure the coop is fully enclosed with no gaps larger than half an inch. Use latches and locks that are raccoon-proof, as raccoons are known for their dexterity and ability to open simple latches.

Motion-Activated Lights and Alarms: Installing motion-activated lights or alarms around the coop can help deter nocturnal predators like raccoons, foxes, and coyotes. When the light or sound activates, it may scare the predator away.

Hawks, Owls, and Other Birds of Prey

Hawks, owls, and other birds of prey can swoop down and grab chickens, especially younger or smaller birds. Protecting your flock from aerial predators is particularly important if they spend time in an open run.

Covering the Run: The best way to prevent aerial attacks is by covering the run with netting, hardware cloth, or wire mesh. A fully enclosed run will protect your chickens from hawks and owls while allowing them to enjoy time outside.

Fake Predators: Some chicken keepers use fake owls or hawks as a deterrent. While these can work for a short time, birds of prey often get used to them, so this method is best used alongside other protections.

String or Wire Grids: Another method to protect against hawks is to run string or wire across the top of the run in a grid pattern. This creates a visual barrier that deters hawks from diving in but still allows sunlight and airflow.

Neighborhood Dogs and Cats

Neighborhood dogs and cats can pose a serious threat to backyard chickens, especially if they're not supervised. Even well-meaning pets can cause stress or injury to chickens.

Fencing: Ensure that your fencing is tall and sturdy enough to keep dogs and cats out. A fence at least 6 feet tall will deter most dogs, and a fully enclosed run will keep both dogs and cats away from your flock.

Supervised Free-Ranging: If you allow your chickens to free-range, it is important to supervise them if you know there are neighborhood dogs or cats around. Consider using portable fencing or chicken tractors to allow your chickens some freedom while still keeping them protected.

Predator-Proof Coop Design

Designing a predator-proof coop from the start can save you time and heartache down the road. By incorporating a few key design features, you can ensure that your coop is secure and your flock is safe from attacks.

Strong, Secure Fencing: The first line of defense against predators is strong fencing around the run. Use heavy-duty materials like hardware cloth or welded wire, and make sure the fence is buried or reinforced to prevent digging. The mesh size should be no larger than ½ inch to prevent smaller predators from squeezing through.

Buried Fencing: Predators like raccoons, foxes, and coyotes are known to dig under fences to get into chicken runs. Burying hardware cloth or wire fencing at least 12 inches underground or creating an outward-facing "predator apron" can prevent them from tunneling in.

Electric Fencing: For added protection, consider installing electric fencing around the perimeter of your coop and run. Electric fences provide an extra layer of security by delivering a mild shock to any predator that tries to breach the fence.

Secure Coop Doors and Latches

Predators like raccoons are known for their ability to open simple latches and squeeze through small gaps. Make sure your coop doors and windows have secure latches that lock and can't be easily pried open.

Automatic Coop Doors: Automatic coop doors that close at dusk and open at dawn are a great way to ensure your chickens are safely locked inside every night. These doors are often predator-proof and can help reduce the risk of nighttime attacks.

Raccoon-Proof Latches: Use double-action latches or padlocks that require two movements to open. Raccoons can easily manipulate simple latches, so securing doors and windows with complex locks will help keep them out.

Elevated or Raised Coops

Building your coop on stilts or raising it off the ground provides an extra layer of protection against predators. An elevated coop makes it more difficult for predators to reach the chickens and eliminates hiding places underneath the coop where predators might burrow or nest.

Advantages of Raised Coops: Elevating the coop keeps the structure drier, reduces rodent access, and provides extra ventilation, all while helping to protect your flock from ground-based predators.

Natural and Non-Toxic Pest and Predator Deterrents

In addition to physical barriers, there are natural and non-toxic methods to help deter both pests and predators from your coop.

Essential Oils and Herbs

Certain essential oils and herbs have strong scents that can repel pests and predators. You can use these as part of your coop maintenance routine to keep unwanted visitors away.

Mint and Lavender: Planting mint or lavender around your coop can help repel rodents and flies. These plants are easy to grow and have the added benefit of being safe for chickens.

Essential Oils: Spraying a mixture of water and essential oils like peppermint, citronella, or eucalyptus around the coop can deter pests. These oils are natural repellents for insects and rodents, and they leave a pleasant scent in the coop.

Motion-Activated Lights and Sounds

Predators like raccoons, foxes, and coyotes are often deterred by sudden lights or noises. Installing motion-activated lights or sound alarms around your coop can scare off predators before they get too close.

Solar-Powered Lights: Solar-powered motion-activated lights are a low-maintenance solution that requires no electricity. They can be placed around the perimeter of the coop and will activate when a predator approaches.

Sound Alarms: Motion-activated sound alarms or sprinklers can startle predators and deter them from coming back. These systems are humane and effective at keeping predators away without harming them.

Troubleshooting Predator and Pest Issues

Despite your best efforts, you may still encounter pest or predator problems. Here are some tips for troubleshooting common issues.

Pests in the Coop

If you notice an increase in pests like mites, flies, or rodents, it is important to act quickly to prevent an infestation from spreading.

Solution: Regularly inspect the coop for signs of pests and keep the area clean and dry. If you find pests, use natural treatments like diatomaceous earth or essential oils to control them. For more severe infestations, consider using commercial pest control products designed for poultry.

Predator Breaches

If a predator manages to breach your coop or run, you will need to identify the weak point and reinforce it immediately.

Solution: Walk around the coop and run, inspecting for any gaps, holes, or signs of digging. Check fences, doors, and windows for weaknesses. Reinforce vulnerable areas with hardware cloth, locks, or additional fencing to prevent future breaches.

By implementing a combination of strong coop design, physical barriers, and natural deterrents, you can protect your flock from pests and predators with minimal maintenance. Keeping your chickens safe and your coop secure is a key part of maintaining a self-sustaining system that runs smoothly with less intervention.

Simplified Health Checks and Biosecurity

Keeping chickens healthy can seem like a daunting task, but by establishing a low-effort routine, it is possible to maintain a thriving flock without constant intervention. Simplified health checks and a strong biosecurity strategy can make all the difference in ensuring the well-being of your chickens while maintaining the self-sufficient, low-maintenance system you have set up.

Establishing a Low-Effort Routine for Monitoring Chicken Health
The key to maintaining a healthy flock is consistency. By incorporating quick, daily health checks into your routine, you can identify potential problems early before they escalate. You do not need to be a veterinarian to spot when something's off. Here's a simplified approach to daily and weekly health checks:

Daily Observations

A lot can be learned just by watching your chickens go about their day. Are they moving around energetically? Do they have good posture? Chickens that are lethargic, standing hunched up, or isolating themselves from the group may be unwell. Check for consistent eating and drinking habits, and listen for any unusual noises like coughing, sneezing, or wheezing, which can signal respiratory issues.

Feather and Skin Condition

While you may not handle your chickens every day, take note of their feathers during your daily checks. Bright, shiny feathers are a good indicator of overall health, while dull, ruffled feathers may indicate parasites, malnutrition, or illness. A quick visual check of their eyes, comb, and wattles (the fleshy parts on their head and neck) should show bright colors and smooth surfaces. Any paleness, swelling, or discharge could be a sign of illness.

Weekly Checkups

Once a week, take the time to pick up a few chickens for a more thorough check. Look closely at their vent (the area under the tail where they lay eggs and excrete waste) for signs of swelling, redness, or discharge, which could indicate issues like egg binding or infections. Feel their keel bone (the ridge of bone in the middle of their chest) to ensure they have enough body fat—too thin, and they might not be eating enough, while too fat can lead to other health issues.

Identifying Common Chicken Diseases and Preventive Measures
Chickens are generally hardy animals, but they can still fall victim to common ailments. Knowing the most likely threats and how to prevent them will make your flock's health checks much more manageable. Here are a few common diseases and some straightforward preventive measures:

Coccidiosis

This parasitic disease affects the intestinal tract and is particularly dangerous for young chicks. Symptoms include diarrhea (sometimes bloody), weight loss, and lethargy. Prevention is key, and can be achieved by keeping the coop dry and clean, as the parasite thrives in damp, dirty environments. You can also purchase medicated feed for young chicks, which contains a small dose of medication to help prevent the disease from taking hold.

Marek's Disease: A viral infection that affects the nervous system, leading to paralysis and often death. There's no cure, but vaccination is an effective preventive measure. If you are raising chicks from a hatchery, be sure to request vaccinated chicks to lower your flock's risk. Keep in mind that Marek's disease is spread through dander, so maintaining a clean coop is critical in keeping this virus at bay.

Avian Influenza

While relatively rare in backyard flocks, avian flu can be devastating. Symptoms include respiratory distress, coughing, sneezing, and swelling of the comb and wattles. The best prevention is good biosecurity (more on that later) and ensuring your flock does not come into contact with wild birds, which can carry the disease.

Bumblefoot

This bacterial infection typically enters through cuts or scrapes on a chicken's feet. Left untreated, it can lead to swelling, lameness, and severe infection. You can prevent bumblefoot by keeping your coop clean and ensuring roosts and perches are smooth and not causing abrasions. If you notice a chicken limping, check for swelling or dark scabs on the bottom of the feet, and treat promptly with cleaning and bandaging.

Reducing the Spread of Illness in a Self-Running System
In any flock, there's always the potential for illness to spread quickly due to the close quarters in which chickens live. However, with a few simple strategies, you can minimize the risk of disease transmission in your self-sustaining system:

Maintain Clean Water and Feed: Clean waterers and feeders daily. Chickens can be messy, and contaminated water or feed can spread illness through the flock. Make sure their water is always fresh and that feeders are kept free from moldy or stale food.

Dry Bedding and Coop Hygiene: Moisture in the coop is a breeding ground for bacteria and parasites. Using deep litter bedding that you turn regularly can help absorb moisture and reduce the spread of disease. Make sure you clean out wet or soiled areas promptly, and replace bedding as needed.

Space and Ventilation: Overcrowding is one of the leading causes of disease outbreaks in chicken flocks. Ensure that each chicken has enough space to roam and that your coop has proper ventilation to keep air flowing and reduce moisture buildup. This will help lower the risk of respiratory diseases and parasites like mites and lice.

Monitor Flock Behavior

Pay attention to changes in behavior. Chickens are social animals, so if one starts to isolate itself, it could be an early sign of illness. Catching issues early can make treatment easier and prevent the spread of sickness to the rest of the flock.

Quarantine Systems and Protecting Your Flock from Outside Threats.

One of the most important elements of biosecurity is managing the introduction of new birds and protecting your flock from outside threats. Whether you are expanding your flock or simply taking precautions against wild animals, proper quarantine and biosecurity measures are essential for preventing disease.

Quarantine New Birds

If you are adding new chickens to your flock, always quarantine them for at least two weeks before introducing them to the others. This gives you time to observe the new birds for signs of illness, such as respiratory issues, lethargy, or diarrhea, without risking the health of your established flock. During this period, keep the new birds in a separate area and use different equipment to care for them. Avoid cross-contamination by washing your hands and changing boots or clothing after handling quarantined birds.

Limit Visitors and Outside Contamination

Chickens can catch diseases from other animals or people who have been in contact with other flocks. Limit access to your chickens from visitors, especially those who keep birds of their own. Wild birds can also carry diseases that threaten your chickens, so keep feeders and waterers inside the coop or covered areas to minimize contact. Also, consider using netting or fencing to reduce the chance of wild birds entering the run.

Footbaths and Protective Gear

For extra biosecurity, place a footbath outside your coop, filled with a disinfectant solution. Stepping through the footbath before entering the coop helps reduce the chances of bringing in bacteria or viruses on your shoes. If you regularly work with other livestock or poultry, consider designating specific clothing and boots for working with your chickens.

By establishing a simple, low-effort routine for health checks and maintaining strong biosecurity measures, you can reduce stress and enjoy the benefits of a self-sufficient chicken system. Keeping an eye on your flock's health does not have to be overwhelming—once you know what to look for and how to act preventively, the process becomes second nature. In the next chapter, we'll dive into how to turn one of the messiest parts of chicken keeping—waste—into a valuable resource for your garden.

Composting Chicken Waste – Turning Mess into Gold

Introduction to Composting and Its Benefits

Keeping chickens can be more than just a source of fresh eggs and companionship—it can also be an environmentally friendly way to enrich your garden. Chicken manure is a powerful organic fertilizer, rich in nitrogen, phosphorus, and potassium, which are essential nutrients for healthy plant growth. However, using manure directly on plants can cause problems due to its high nitrogen content, making composting an essential step.

Composting transforms chicken waste into a nutrient-rich soil amendment that can boost your garden's health while reducing waste from the coop. Some key benefits include:

Reduces Waste: You turn what would otherwise be a nuisance into a valuable resource.

Improves Soil Health: Compost enhances soil structure, moisture retention, and fertility.

Cost-Effective: It eliminates the need for commercial fertilizers.

Environmentally Friendly: Composting is a natural, sustainable way to recycle waste from your coop.

Setting Up a Simple Chicken Waste Composting System

Starting a composting system for your chicken manure doesn't require an elaborate setup. A simple approach using readily available materials can yield great results. The basic elements you will need include a compost bin, chicken manure, and bedding (straw or wood shavings).

Choosing the Right Compost Bin

The first step is selecting a compost bin that fits your space and lifestyle. Here are some popular options:

Open Pile: Designate a corner of your yard to pile up the chicken manure and bedding. This option requires little setup but does need frequent turning.
Compost Bin: Using a compost bin helps contain the pile, reducing the chances of attracting pests. You can buy a commercial bin or build your own from wood or wire mesh.
Compost Tumbler: Tumblers are ideal for small spaces and make it easy to aerate the compost by turning it regularly. They can speed up the composting process significantly.
Choose a location that's convenient but out of direct sunlight, as too much heat can dry out the pile and slow down decomposition.

Collecting Chicken Waste and Bedding

Chicken waste alone is too rich in nitrogen to compost efficiently. It needs to be balanced with carbon-rich materials like bedding (wood shavings, straw, or leaves). Each time you clean the coop, collect both manure and bedding to create the right balance for your compost pile.

A good rule of thumb is to aim for a 2:1 ratio—two parts bedding to one part manure. This balance ensures your compost will break down without generating unpleasant odors or attracting pests.

Balancing Nitrogen and Carbon (Greens and Browns)

The key to effective composting is maintaining the right balance between nitrogen-rich "greens" (manure) and carbon-rich "browns" (bedding). Here's a simple way to think about it:

If the pile is too wet or smells like ammonia, it has too much nitrogen. Add more bedding (browns).

If the pile is too dry or decomposing slowly, it needs more moisture or nitrogen. Add water or more manure (greens).
Aim for a texture similar to a damp sponge—moist but not soggy.

Aerating the Compost Pile

Composting is an aerobic process, which means the pile needs oxygen to decompose effectively. To ensure your compost pile breaks down evenly, turn it with a pitchfork every 2-3 weeks. For compost tumblers, rotating the bin every few days speeds up the process.

Regular turning introduces oxygen to the microorganisms that break down the material, ensuring faster and more complete composting.

Letting the Compost Cure

After about 4-6 months of active composting, your pile should turn into a dark, crumbly material that smells sweet and earthy. While it may look like it is ready to use, the compost still needs to cure. Let it sit undisturbed for 1-2 months to allow any remaining material to fully break down.

Once fully cured, the compost is safe to use in your garden, providing your plants with a steady supply of nutrients.

Using Chicken Manure Compost in Your Garden

Composted chicken manure is one of the most effective soil amendments you can use. Here's how to incorporate it into your gardening routine:

Amend Garden Beds: Before planting, spread a 1-2 inch layer of compost over your garden beds and work it into the top few inches of soil. This improves the soil's structure and boosts fertility.
Top-Dressing: Use compost as a slow-release fertilizer by applying a thin layer around the base of established plants. The nutrients will gradually seep into the soil as the compost breaks down.
Mulch: Compost can also serve as a mulch, helping to retain moisture, suppress weeds, and feed plants over time.

Tips for Success

Do not overdo it: Chicken manure compost is potent, so less is more. A thin layer of compost will go a long way.
Soil Testing: Consider testing your soil before applying compost to avoid over-fertilizing, especially if your soil is already nutrient-rich.
Compost Year-Round: Keep your composting system running throughout the year to ensure a steady supply of organic matter for your garden.

Troubleshooting Common Composting Problems

While composting is a straightforward process, you might run into a few snags along the way. Here are some common issues and their solutions:

Bad Odor: If your compost pile smells foul, it may have too much nitrogen or moisture. Add more bedding and turn the pile to increase aeration.
Slow Decomposition: If your compost isn't breaking down, it may be too dry or lack sufficient nitrogen. Add water or fresh manure to speed up the process.
Pests or Flies: Cover fresh manure with bedding immediately after adding it to the pile to avoid attracting pests. A well-maintained pile should not smell or attract unwanted critters.

Integrating Composting into Your Self-Running Coop

For those seeking a low-maintenance, self-sustaining chicken coop, integrating composting into your routine is a natural fit. Here are a few tips to make it even easier:

The Deep Litter Method: This method allows manure and bedding to compost in place right on the coop floor. Simply add fresh bedding as needed, and only clean out the coop a few times a year.

The materials break down gradually, creating a compost-like material you can use in your garden.

Automated Solutions: Consider investing in automatic waterers and feeders to reduce time spent on daily tasks. Paired with a composting system, your coop can practically run itself.

Final Thoughts on Composting Chicken Waste: Turning chicken waste into compost is one of the best ways to close the loop on your self-sufficient chicken coop system. Not only does it reduce waste and keep your coop cleaner, but it also produces a powerful, organic fertilizer that can dramatically improve your garden's productivity.

With a little knowledge and regular maintenance, composting becomes a seamless part of your routine, providing endless benefits to both your chickens and your plants. As you implement composting, you will find that your self-running coop isn't just efficient for your flock—it is a key part of a thriving, eco-friendly homestead.

Raising Chickens for Eggs – Laying Hens and Care

Egg production is one of the main reasons people start raising chickens. However, keeping hens productive while minimizing your involvement is key to running a successful self-sustaining chicken coop. In this chapter, we'll cover the essentials of raising laying hens, including selecting breeds, maintaining egg production, and long-term care practices that ensure your hens stay healthy and productive without requiring constant attention.

Selecting Breeds That Thrive in Low-Maintenance Environments

Not all chicken breeds are created equal when it comes to egg production and ease of care. Some hens are hardy, efficient layers that do well in a self-sustaining system, while others may require more hands-on management. When choosing breeds for a low-maintenance coop, consider the following factors:

Egg Production Rate

Different breeds lay at different rates. For a productive flock with minimal intervention, choose breeds known for high egg output. Here are a few top choices:

Rhode Island Red: Known for their resilience and high egg production, these birds lay around 250-300 eggs per year.

Leghorn: These lightweight hens are prolific layers, producing up to 280-320 eggs annually.

Sussex: A dual-purpose breed that offers both meat and eggs, Sussex hens lay about 250 eggs per year.

Australorp: Famous for holding the world record for egg production, these hardy birds produce 250-300 eggs annually.

Temperament and Behavior

Calm, easygoing hens are easier to manage in a low-maintenance system. Look for breeds that are:

Friendly and docile: Breeds like Buff Orpingtons or Plymouth Rocks are known for their gentle nature and are easy to handle.

Good foragers: If you plan to allow your flock to free-range, breeds that are natural foragers (like Wyandottes) can help reduce your feed costs by foraging for food on their own.

Climate Tolerance

Choose breeds that can handle your local climate with minimal intervention:

For cold climates, consider breeds with small combs and plenty of feathering to reduce the risk of frostbite (e.g., Chanteclers or Brahmas).

For hot climates, lightweight breeds with large combs (such as Leghorns) are better suited to dissipating heat.

Best Practices for Keeping Egg Production High with Minimal Involvement

Once you've selected the right breeds, the next step is ensuring your hens maintain consistent egg production without requiring constant oversight. While egg production can fluctuate based on several factors, these tips can help keep your hens laying regularly:

Provide a Balanced Diet

A hen's diet plays a major role in egg production. To keep production high:

Ensure your chickens always have access to layer feed that is high in calcium and protein, which are essential for egg formation. Allow free-ranging or provide supplemental greens, insects, and grains to encourage natural foraging behavior. Crushed oyster shells should be available as a source of calcium to ensure strong eggshells.

Maintain a Comfortable Environment

Stress is a major factor that can reduce egg production. Keeping your coop comfortable and clean helps hens stay productive:

Temperature control: Ensure proper ventilation in summer and adequate insulation in winter.

Low-stress nesting areas: Provide enough nesting boxes (one box for every 3-4 hens) in quiet, dimly lit areas to reduce competition and stress.

Consistent light: Hens need about 14-16 hours of light daily to maintain steady egg production. In the winter months, using timed artificial lighting can keep their laying cycle consistent.

Water and Hydration

A consistent supply of fresh, clean water is crucial for egg production: Invest in automated watering systems to ensure hens always have access to water. In winter, consider heated waterers to prevent freezing and reduce the need for manual checks.

Egg Collection Routine

Regular egg collection not only keeps the coop clean but also encourages consistent laying. Leaving too many eggs in the nesting boxes can lead to broken eggs or encourage broodiness (when hens stop laying and try to hatch eggs).

Collect eggs at least once a day, preferably twice (morning and afternoon).
Regular egg collection also prevents egg-eating behavior in the flock.

Long-Term Care for Hens: Balancing Longevity and Productivity

Egg-laying hens have productive and non-productive phases throughout their lives. Understanding these cycles helps manage the flock for both longevity and continued production.

Molting: A Natural Pause in Egg Production

Molting is the process where chickens shed their old feathers and grow new ones, which typically happens once a year. During this time, hens stop laying eggs because their bodies are focused on feather regrowth rather than egg production. This process usually lasts 8-12 weeks, but can extend longer depending on the hen's health and environment.

To support hens during molting:

Increase their protein intake by offering protein-rich foods like mealworms, scrambled eggs, or high-protein feeds.
Avoid stressing the birds during this time, as stress can prolong the molting process.

Keep the coop warm and free from drafts, especially if molting occurs during colder months.

Managing Older Hens

As hens age, their egg production naturally slows down. By year 3, many hens will begin laying fewer eggs, and by year 5, most will produce significantly fewer or stop laying altogether. Here are a few strategies for managing older hens:

Retirement plan: Some chicken keepers allow older hens to live out their natural lives as companions.

Culling or rehoming: If maintaining a productive flock is a priority, some chicken keepers choose to cull or rehome older hens that no longer produce.

Encouraging Longevity and Health

To keep hens healthy and productive for as long as possible, practice preventative care:

Regularly check for signs of illness, such as weight loss, lethargy, or changes in behavior. Ensure hens have access to dust baths to help control external parasites like lice and mites.

Provide ample space for the birds to roam and avoid overcrowding, as it can lead to stress, disease, and lower egg production.

Managing Broodiness and Molting

Broodiness is when a hen stops laying eggs and attempts to hatch them, even if they're infertile. While this is a natural behavior, it can interrupt egg production in a self-running system. Managing broodiness effectively can help maintain productivity with minimal effort.

Recognizing Broody Behavior

A broody hen will:

Spend most of her time sitting on the nest, often growling or puffing up when disturbed.

Pluck feathers from her breast to create a bare patch for better contact with the eggs.

Stop laying eggs and focus solely on sitting on the nest.

Breaking Broodiness

If you do not plan to let a hen hatch chicks, breaking her broodiness will get her back to laying eggs faster. Here are a few methods:

Remove her from the nest: Gently take her out of the nesting box and close off access.

Cool her down: Broody hens tend to be warm, so placing her in a wire-bottomed cage where cool air can circulate underneath may help break the behavior.

Limit access to nests: If multiple hens are broody, close off the nesting area entirely for a few days.

Minimizing Molting Disruptions

Molting can not be avoided, but minimizing stress during this period can help the process go smoothly:

Keep the flock's diet high in protein to support feather regrowth. Reduce handling or sudden changes in the environment to avoid additional stress.

A Low-Maintenance Approach to Egg Production

Raising laying hens for consistent egg production does not have to be labor-intensive. With the right breed selection, proper care routines, and smart systems in place, your hens can continue producing eggs with minimal daily effort on your part.

By understanding the natural cycles of hens—like molting and broodiness—and providing the right balance of nutrition, hydration, and environmental comfort, you can enjoy a productive flock with very little ongoing maintenance. In the next chapter, we will explore how to use natural solutions to maintain a healthy, happy coop environment.

Natural Solutions for a Happy Coop

Maintaining a healthy and happy chicken coop does not require harsh chemicals or complicated systems. By incorporating natural solutions, you can create a sustainable environment that benefits both your chickens and your homestead. In this chapter, we will explore how to use herbs, plants, and other natural remedies to keep your coop healthy, control pests, and improve the overall well-being of your flock.

Incorporating Herbs, Plants, and Natural Remedies in Your Coop

Herbs and plants are a natural, low-maintenance way to improve your chicken coop's cleanliness and support your chickens' health. By strategically growing and placing certain plants around the coop, you can control pests, reduce odors, and even enhance your chickens' immunity.

The Benefits of Using Herbs in Your Coop

Pest control: Some herbs act as natural repellents for common pests like mites, lice, and flies.

Odor reduction: Aromatic herbs can help neutralize ammonia and other odors in the coop.

Immune support: Certain herbs, when eaten or used in dust baths, can boost chickens' immune systems.

Calming effects: Some herbs can have a soothing effect on hens, reducing stress and encouraging egg laying.

Best Herbs to Grow for Chickens

Here are some popular herbs that can be grown near the coop or added to nesting boxes for natural health benefits:

Lavender: Lavender's scent helps calm chickens, reducing stress and promoting relaxation. It can also repel pests like flies and mosquitoes.

Mint: Mint repels rodents, insects, and parasites. You can grow it around the coop or dry the leaves to sprinkle inside.

Sage: Known for its antibacterial properties, sage helps boost immunity and fights off infections in your flock.

Rosemary: A natural insect repellent, rosemary is great for keeping mites and lice at bay. Its strong aroma also helps neutralize coop odors.

Thyme: Thyme has strong antimicrobial properties and can be added to chicken feed or bedding to fight off bacteria and support respiratory health.

Oregano: Oregano is a powerful natural antibiotic that supports the immune system and is often added to chickens' water or feed to prevent illness.

How to Use Herbs in the Coop

There are several ways to incorporate herbs into your coop:

In nesting boxes: Add dried herbs to nesting boxes to repel pests and provide a calming environment for laying hens.

As bedding additives: Mix herbs with straw or wood shavings in your bedding to fight odors and deter mites.

In dust baths: Create a dust bath area with herbs like lavender, rosemary, or mint to help chickens control external parasites naturally.

Around the coop: Plant herbs around the perimeter of your coop to create a natural barrier against insects and rodents.

Using Plants for Pest Control, Odor Reduction, and Chicken Well-Being

Beyond herbs, certain plants can improve the environment around your coop and run. Choosing the right plants can help you maintain a healthy, low-maintenance setup by managing pests, improving the chickens' overall environment, and reducing unpleasant odors.

Plants that Repel Pests

Insects and parasites like flies, mites, and lice can cause stress and health issues for chickens. By planting natural pest-repellent plants near the coop, you can help keep these nuisances at bay:

Marigolds: These bright flowers repel a wide variety of insects, including mosquitoes, flies, and beetles. They are also safe for chickens to peck at.

Basil: Basil helps repel flies and mosquitoes and can be grown easily in pots around the coop.

Catnip: Catnip is highly effective at repelling mosquitoes and flies, while being harmless to chickens.

Plants that Improve Coop Odor

Ammonia buildup from chicken manure can lead to unpleasant smells in the coop, but planting fragrant, air-purifying plants can help:

Lemon Balm: This fast-growing plant has a citrusy smell that helps mask coop odors and is safe for chickens to eat.

Eucalyptus: The oil from eucalyptus leaves is known for its strong, refreshing scent. It helps reduce foul smells and has antibacterial properties.

Plants that Support Chicken Health

Some plants can also have direct health benefits for chickens, particularly in improving digestion and boosting immunity:

Dandelions: Dandelions are packed with vitamins and minerals, and they can be a great supplement to your chickens' diet. They support digestion and boost overall health.

Comfrey: Comfrey is rich in protein, and its leaves can be fed to chickens as a nutritious snack. It also has anti-inflammatory properties that support healing.

Creating a Coop-Friendly Garden

Building a small coop-friendly garden around your chicken run not only improves the environment but also reduces your workload. Chickens will naturally forage for insects and plants, so growing edible plants around the run can help provide them with variety in their diet while you enjoy the benefits of fewer pests and odors.

How to Plant a Coop-Friendly Garden that Benefits Both Chickens and Owners

A coop-friendly garden can be functional and beautiful, providing both practical benefits for chicken keeping and aesthetic value for the homestead. Design a garden that supports both your chickens and your lifestyle.

Designing the Garden

When planning a coop-friendly garden, think about the placement of plants to maximize benefits:

Border plants: Use pest-repelling plants like marigolds, basil, and catnip along the borders of the coop and run to create a natural barrier against insects.

Shade plants: In hot climates, plant fast-growing vines or shrubs around the coop to provide extra shade. Plants like grapevines, sunflowers, or bamboo can offer much-needed relief from the summer heat.

Chicken-friendly snacks: Grow chicken-safe plants inside the run, such as clover, dandelions, and comfrey, to give your flock access to fresh greens.

Protecting Delicate Plants from Chickens

Chickens love to scratch and peck, which can be destructive to delicate plants. Here's how to protect your garden while still giving your chickens access to beneficial plants:

Use raised beds or fencing around delicate plants.
Plant hardier, chicken-safe plants like comfrey or clover inside the run, as these are more resistant to pecking and scratching.
Grow more fragile herbs and flowers outside the run, ensuring they remain accessible to you but protected from curious chickens.

Creating Natural Shade and Shelter

If your chickens spend a lot of time outdoors, providing natural shade and shelter can reduce stress, particularly in the summer. Consider planting trees, bushes, or vines that provide shade around the perimeter of the run. Some options include:

Willow trees: Fast-growing with abundant foliage, willows provide excellent shade.
Grapevines: These vines grow quickly and offer both shade and edible leaves for your chickens.

Sunflowers: Tall and easy to grow, sunflowers create shade while their seeds provide a nutritious treat for chickens.

Encouraging Free-Range Behavior While Maintaining a Low-Maintenance Routine

Free-ranging is one of the best ways to reduce your workload while keeping your chickens healthy and happy. When done right, free-ranging allows your chickens to forage for insects, plants, and other natural foods, reducing your feed costs and improving their well-being. Here's how to encourage free-ranging while maintaining a low-maintenance routine.

Benefits of Free-Range Chickens

Allowing your chickens to free-range offers several advantages:

Reduced feed costs: Free-range chickens will naturally forage for food, reducing the amount of commercial feed you need to provide.
Pest control: Chickens will happily eat insects, slugs, and other pests that could otherwise damage your garden or crops.
Happier chickens: Free-ranging gives chickens the opportunity to express natural behaviors like scratching, foraging, and dust bathing, leading to happier, healthier birds.

Managing Free-Range Chickens

While free-ranging offers many benefits, there are a few things to keep in mind to ensure your chickens remain safe and your routine stays simple:

Perimeter control: If predators are a concern, consider using portable fencing to give your chickens access to safe foraging areas while keeping them protected.
Automated systems: Set up automatic doors on the coop that open at dawn and close at dusk to let your chickens out in the morning and secure them at night without you having to monitor their routine.

Rotational grazing: If you have the space, rotate your chickens between different areas of the yard or garden. This reduces overgrazing and allows the soil in each area to recover.

Free-Range Safety Considerations

While free-ranging can reduce your workload, safety is a priority. Here are some ways to keep your chickens safe while they roam:

Predator-proof fencing: Ensure the free-range area is enclosed or monitored to keep predators like hawks, foxes, or raccoons at bay.
Safe hiding spots: Provide bushes, tall plants, or chicken tunnels where your birds can quickly hide from predators if necessary.

Water stations: Keep waterers in shaded areas throughout the free-range zone to ensure chickens stay hydrated without needing to return to the coop constantly.

A Natural Approach to Chicken Coop Care

Incorporating natural solutions into your chicken coop can improve the health and well-being of your flock, reduce pests, and enhance the overall environment with minimal effort. By using herbs and plants strategically, you can naturally control pests, reduce odors, and create a pleasant, low-maintenance coop.

Free-ranging chickens also play an important role in a self-running system, reducing feed costs and offering natural pest control. When designed thoughtfully, a coop-friendly garden and free-range setup can make your chicken-keeping routine easier and more rewarding.

In the next chapter, we'll explore seasonal maintenance, offering tips and strategies for simplifying tasks throughout the year while keeping your coop running smoothly.

Seasonal Coop Maintenance Made Simple

One of the keys to maintaining a self-running chicken coop is understanding how to adapt your management practices to the changing seasons. Each season presents its own challenges, from the sweltering heat of summer to the freezing temperatures of winter. By breaking down coop maintenance into seasonal tasks, you can stay ahead of potential problems and keep your chickens comfortable and healthy year-round with minimal effort.

What to Expect and Plan for During Each Season

The seasonal changes affect your chickens in various ways, from their water needs to the amount of light they receive, and each season requires slight adjustments to your management routine. This section outlines what to expect during the year and how to plan for it.

Spring: Preparing for the Year Ahead

Spring is an ideal time to freshen up your coop and make sure everything is in working order after winter. As temperatures warm up, your chickens will start laying more eggs, and pests like mites and flies may become more active.

Tasks for spring include:

Deep cleaning the coop: After a long winter, give the coop a thorough clean by removing all old bedding, sanitizing the floors, and replacing it with fresh materials.

Check the structure: Inspect the coop for any damage caused by winter weather, such as leaks, rust, or drafts, and make any necessary repairs.

Replenish bedding: Add fresh bedding materials to the coop and nesting boxes. Consider adding herbs like mint or lavender to help repel pests as the weather warms up.

Pest control: Start using diatomaceous earth or natural repellents in the coop to control mites, lice, and flies before they become a problem.

Prepare for egg production: Clean and refill the nesting boxes as hens ramp up their laying with the longer daylight hours.

Summer: Managing Heat and Ventilation

Summer presents the challenge of keeping your chickens cool, hydrated, and protected from the heat. Overheating can cause stress and reduce egg production, so it is important to make adjustments to your coop and run during this time.

Key tasks for summer include:

Maximize ventilation: Ensure that your coop has good airflow to prevent overheating. Open windows or vents to create cross-breezes, or install solar-powered fans for additional cooling.

Provide shade: Use shade cloth, trees, or other plants to create shaded areas in the run where chickens can escape the sun. Chickens will seek out cooler spots, especially in the heat of the day.

Keep water cool: Check water stations frequently and refill them with cool, clean water. In extreme heat, consider adding ice cubes to the water or using automatic waterers with a consistent flow to keep water from getting too warm.

Add dust baths: Dust bathing is how chickens cool down and keep external parasites away. Ensure they have access to a dust bath area in a shady spot.

Reduce bedding thickness: In the summer, using thinner layers of bedding can help with airflow and reduce heat buildup inside the coop.

Fall: Preparing for Winter

Fall is the perfect time to prepare for the cold months ahead. Your hens will likely slow down or stop laying eggs as the days get shorter, and their bodies will focus on growing new feathers during their annual molt. Use this transitional time to prepare your coop for winter.

Essential tasks for fall include:

Insulate the coop: Check for any drafts or leaks in the coop and patch them up. While chickens tolerate cold better than heat, drafts can lead to frostbite or illness. Consider adding extra straw or hay bales around the outside of the coop for added insulation.

Deep litter method: If you are using the deep litter method, now is the time to start building up the layers of bedding. The decomposing materials will generate some warmth during the winter months.

Install heated waterers: Ensure you have a reliable way to keep water from freezing in winter. Heated waterers or water heater bases are a great investment and can prevent the need to constantly check and replace frozen water.

Prepare for molting: As your chickens lose feathers during the fall molt, provide extra protein-rich feed to help them regrow feathers and stay warm as the temperature drops.

Install artificial lighting: If you want to maintain egg production through the winter, consider setting up timed artificial lights to extend daylight hours. Aim for 14-16 hours of light per day to keep hens laying.

Winter: Keeping Chickens Warm and Water Flowing

Winter is the most challenging season for chicken keepers due to cold temperatures, shorter days, and the potential for frozen water. However, chickens are surprisingly hardy, and with a few preparations, you can keep your flock comfortable even during the coldest months.

Key tasks for winter include:

Insulate but ventilate: While it is important to keep the coop warm, make sure you do not block off ventilation entirely. Proper airflow is needed to prevent moisture buildup, which can lead to frostbite. Ensure the coop stays draft-free but still allows air circulation.

Use supplemental heating cautiously: Chickens can usually withstand the cold without additional heat, but in extreme conditions, you might consider using a coop heater or heat lamps. Be cautious of fire hazards, and only use heating devices if absolutely necessary.

Monitor waterers: Frozen water is one of the biggest issues in winter. Use heated water bowls or change water frequently to ensure your chickens always have access to clean, unfrozen water.

Increase calorie intake: Chickens need extra energy to stay warm, so consider adding scratch grains or extra feed in the evening. The process of digestion helps generate heat in their bodies.

Seasonal Cleaning Checklists and Efficiency Hacks

Each season comes with its own set of cleaning tasks, but by breaking down these tasks into manageable checklists, you can streamline your coop maintenance routine. Here are seasonal cleaning checklists and tips for staying efficient.

Spring Cleaning Checklist

Remove all old bedding and sanitize the coop floors and walls. Clean nesting boxes and refresh with new bedding and herbs. Inspect and repair any damage to the coop structure. Clean and refill feeders and waterers.

Add diatomaceous earth in corners and crevices to prevent mites. Efficiency Tip: Schedule your spring cleaning on a warm, dry day so that any cleaning solutions dry quickly, and you can air out the coop easily.

Summer Cleaning Checklist

Lightly rake and turn over bedding to prevent ammonia buildup. Wipe down waterers and feeders weekly to prevent algae growth. Remove droppings from nesting boxes daily to reduce flies and odors.

Keep windows and vents clear of debris to ensure proper airflow.

Efficiency Tip: Use sand or pine shavings in the run during summer, as they help reduce odors and dry out quickly, making them easier to manage.

Fall Cleaning Checklist

Deep clean the coop before winter sets in, removing all bedding and sanitizing the interior. Inspect the roof and walls for leaks or gaps and seal them.

Start building up layers of bedding if using the deep litter method. Clean and check waterers to ensure they're ready for winter conditions.

Efficiency Tip: Perform major repairs and inspections in the fall before freezing temperatures set in to avoid working in cold, harsh weather.

Winter Cleaning Checklist

Spot clean droppings to prevent moisture buildup in the coop.

Check bedding for wet spots and add fresh layers as needed.

Clear pathways and entrances of snow or ice to ensure easy access for both you and your chickens.

Efficiency Tip: In winter, cleaning tasks should focus more on spot cleaning and managing moisture rather than deep cleaning. Keeping the coop dry and well-insulated is more important than completely changing bedding during cold weather.

Preparing for Winter: Insulation, Snow Removal, and Keeping Water from Freezing

Winter prep is critical to ensuring your chickens stay healthy and comfortable. Here are some specific strategies for getting your coop winter-ready:

Insulating the Coop

Draft-free environment: Seal any cracks or holes where cold air can enter, but avoid sealing off all ventilation, as fresh air is still essential.

Deep litter method: As the bedding layers decompose, they generate heat, which helps keep the coop warmer during winter. This natural insulation is a simple, low-maintenance way to keep your chickens warm.

Snow Removal and Access

Clear pathways: Keep the paths around the coop free of snow, so you and your chickens can move easily. Use gravel or wood chips to provide better traction in icy conditions.

Snow load: Check the coop roof regularly to prevent heavy snow accumulation, which can cause structural damage or leaks.

Keeping Water from Freezing

Heated waterers: These are the easiest way to ensure water stays thawed in cold weather. Heated waterers or bases prevent freezing without needing constant manual refills.

Insulated waterers: If you do not have access to power for heated waterers, consider using insulated water containers or adding a ping pong ball to the water, which moves with the wind and can slow down freezing.

Heat Management: Shade, Ventilation, and Cooling

Keeping your chickens cool in the summer heat is essential to prevent heat stress, which can lead to health issues or even death in extreme conditions.

Providing Shade

Natural shade: Plant fast-growing trees or vines around the coop to provide natural shade in the hottest part of the day.

Shade cloth: Install shade cloth over part of the run or coop to give your chickens a place to cool off without direct sunlight

Improving Ventilation

Open windows and vents: Ensure there's plenty of airflow by opening windows or vents. If necessary, install solar-powered fans to help circulate air in extreme heat.

Coop design: Ensure your coop's design allows for natural cross-ventilation. Raising the coop off the ground can also help improve air circulation in hotter climates.

Cooling Solutions

Frozen treats: Give your chickens frozen fruits or vegetables like watermelon, cucumbers, or berries to help them cool down.

Misting system: If temperatures are extremely high, consider using a misting system or spraying the coop area lightly with water to reduce the temperature.

Efficient, Low-Maintenance Seasonal Coop Care

Seasonal maintenance does not have to be overwhelming. By planning ahead and implementing simple changes throughout the year, you can keep your coop running smoothly with minimal effort. Preparing for each season, managing potential weather challenges, and using efficient cleaning routines will keep both you and your flock happy year-round.

In the next chapter, we'll explore how to expand your flock while maintaining a self-running system that continues to work efficiently even as your chicken family grows.

Expanding the Flock – Introducing New Birds to a Self-Running System

At some point, most chicken keepers will want to expand their flock—whether by adding new breeds, replacing older hens, or simply increasing egg production. However, introducing new birds to an established system comes with challenges. It is essential to do this carefully to maintain the efficiency of your self-running system and avoid disruptions in the flock. This chapter covers how to introduce new chickens smoothly, monitor pecking order dynamics, and manage the growing flock while maintaining a low-maintenance routine.

The Challenges of Adding New Chickens and How to Minimize Stress

Introducing new chickens into an established flock can be tricky. Chickens are highly territorial and hierarchical animals, and any newcomers will disrupt the existing pecking order. Without proper integration, the process can result in bullying, stress, and injury. The key to a smooth transition is preparation, patience, and careful monitoring.

Understanding the Pecking Order

The pecking order is the natural social hierarchy that chickens establish within a flock. Dominant birds have priority access to food, water, and roosting spots, while lower-ranking birds must wait their turn. When new chickens are introduced, the flock will re-establish this hierarchy, often through pecking and chasing. While some squabbling is normal, it is important to monitor for excessive bullying.

Stress and Health Concerns

Introducing new birds can cause stress for both the new and existing chickens. Stress can result in:

Reduced egg production.

Weakened immune systems, making chickens more susceptible to illness.

Bullying and injuries, especially to younger or smaller birds.

Minimizing stress during this process is key to keeping the entire flock healthy and happy.

Steps to Successfully Introduce New Birds into an Established Coop

To integrate new chickens into your self-running system without causing chaos, follow a gradual and controlled introduction process. This will reduce stress, limit fighting, and ensure that your coop and run continue to function efficiently.

Quarantine First for Health and Safety

Before introducing new birds to your flock, it is essential to quarantine them for at least 2-4 weeks to prevent the spread of disease. Even if new birds appear healthy, they may carry illnesses that could infect your existing flock.

During quarantine:

Keep the new chickens in a separate enclosure far away from your current flock.

Monitor the new birds for signs of illness such as coughing, sneezing, lethargy, or unusual droppings.

Ensure they have fresh water and food and check their health daily.

After the quarantine period, if the new birds show no signs of illness, you can begin the introduction process.

Gradual Introduction Using the "See but Do Not Touch" Method

A slow introduction is the best way to reduce aggression between old and new birds. The "see but do not touch" method allows the chickens to get used to each other without direct contact.

Create a separate but adjacent space: Set up a temporary pen or divide your run so the new chickens can see the established flock, but there's a barrier between them.

Let them interact visually: Allow both groups to observe each other for at least a week. This gives them time to become familiar without physical fighting.

Watch for calm behavior: Once the chickens are used to seeing each other through the barrier and there's less tension, it is a sign that they're ready for closer contact.

Supervised Free-Range Time

Once the chickens have had time to adjust to each other's presence, the next step is to let them interact in a neutral space, such as a large, open area or the free-range zone. This reduces territorial disputes, as no single bird has the "home advantage."

Supervise closely: Keep a watchful eye on the interactions. There will likely be some pecking and chasing as the new pecking order is established, but avoid allowing excessive bullying or injuries.

Provide distractions: Scattering food or treats in the free-range area can help distract the chickens and encourage foraging rather than fighting.

Allow the birds to interact for short periods at first, gradually increasing the time as they adjust to each other.

Full Integration into the Coop

Once the birds have been introduced in a neutral space, it is time to integrate them fully into the coop and run. Here are some tips to make the final transition smooth:

Introduce at night: Adding new chickens to the coop at night can reduce confrontation. Chickens are calmer in the dark and are more likely to accept newcomers when they wake up together in the morning.

Provide extra space: Make sure there are plenty of feeders, waterers, and roosting spots to avoid competition. The more access points there are, the less the new birds will be targeted.

Monitor for aggression: Continue observing the flock for a few days after full integration. Some pecking is normal, but watch for signs of injury or severe bullying, which may require separating the birds again temporarily.

Monitoring the Pecking Order and Preventing Bullying

Once the new birds are fully integrated, the flock will re-establish its social hierarchy. During this time, it is essential to monitor behavior closely to ensure that the pecking order is being established without excessive aggression.

Signs of Normal Pecking Order Behavior

Some pecking and squabbling are natural as chickens establish their ranks. Normal pecking order behavior includes:

Brief squabbles over food or space.

Chasing or light pecking that does not result in injury.

Posturing or feather-fluffing as birds assert dominance.

These behaviors should subside within a week or two as the new pecking order stabilizes.

Signs of Bullying or Aggression

Excessive bullying can cause injury, stress, and even death in extreme cases. Watch for these signs of serious aggression:

Constant pecking or chasing of one specific bird.

Feather loss or visible wounds, especially on the head or back.

Isolation: A bullied chicken may stop eating, drinking, or may hide from the rest of the flock.

If bullying becomes a problem, remove the aggressive bird(s) for a few days to give the lower-ranking chickens a chance to settle in.

Reintroducing the aggressive birds later may help reset their behavior.

Preventing Future Bullying

To prevent bullying and ensure the flock remains peaceful, here are a few strategies:

Provide plenty of space: Overcrowding is one of the leading causes of aggression. Ensure there is enough room for each chicken to eat, drink, and roost without competition.

Offer multiple feeding and water stations: This reduces competition over resources, especially for new or lower-ranking chickens.

Add hiding spots or perches: Creating spaces where bullied birds can escape or find refuge can help reduce stress and give them time to recover.

Maintaining System Efficiency with a Growing Flock

As your flock grows, it is important to adjust your coop and maintenance systems to accommodate more birds while keeping the operation as low-maintenance as possible.

Adjusting the Coop and Run Size

One of the most critical factors in maintaining a healthy, self-sustaining flock is providing enough space. As a general rule:

Each chicken needs at least 4 square feet inside the coop.

In the run, provide 8-10 square feet per chicken to avoid overcrowding.

If you are expanding your flock significantly, consider enlarging the coop or adding a mobile coop to give the birds more space.

Increasing Feed and Water Capacity

As the number of chickens grows, you will need to adjust your feeding and watering systems to ensure all birds have access to food and water without competition.

Automatic feeders and waterers: Install or upgrade your automated feeders and water systems to hold more feed and water, reducing how often you need to refill them.

Multiple stations: In larger flocks, set up multiple feeding and watering stations to minimize crowding and bullying at these critical resources.

Managing Waste and Bedding

More chickens mean more waste, so managing manure and bedding efficiently becomes even more important in a growing flock:

Deep litter method: If you aren't already using it, the deep litter method can reduce the frequency of coop cleaning by allowing bedding to break down naturally over time.

Composting: Composting chicken waste can turn manure into a valuable resource for your garden, and with a larger flock, you will have more material to work with.

Scaling Your Maintenance Routine

As your flock grows, you do not want to increase your workload. Here are a few ways to scale up your flock without increasing maintenance:

Automated systems: Upgrade or add more automation for feeders, waterers, and lighting to reduce daily tasks.

Outsource routine tasks: If you have family members or neighbors who enjoy helping out, consider sharing the workload, especially for larger chores like seasonal cleanouts or health checks.

Expanding the Flock with Minimal Disruption

Introducing new chickens into an established flock does not have to be stressful or complicated. By following a gradual introduction process, monitoring the pecking order, and making simple adjustments to your coop and maintenance systems, you can expand your flock without disrupting the efficiency of your self-running system.

A growing flock can bring new life and energy to your homestead, and with the right planning, it won't add much to your workload. In the next chapter, we'll explore troubleshooting common coop issues and how to fine-tune your systems for maximum efficiency.

Troubleshooting and Fine-Tuning Your Coop

Even in a well-maintained, self-running chicken coop, problems can arise from time to time. Whether it is malfunctioning equipment, pest infestations, or unexpected health issues, troubleshooting is a key skill for any chicken keeper. In this chapter, we'll cover common coop issues, how to diagnose them, and the steps you can take to fine-tune your systems for long-term success.

Common Issues That Arise in Even the Best Setups

No matter how well-planned your chicken coop system is, certain issues can crop up, especially as the seasons change or your flock grows. Being proactive and recognizing these problems early can prevent them from becoming bigger issues down the line.

Watering System Failures

A reliable source of clean water is essential for your flock's health. However, even the best watering systems can fail due to blockages, leaks, or freezing temperatures. Common issues include:

Blocked or clogged waterers: Algae, dirt, or feed debris can clog nipples or other water delivery systems, preventing chickens from accessing water.

Leaks: Automatic waterers or hoses can sometimes develop leaks, leading to water loss or soggy conditions in the coop.

Frozen water: In winter, even well-insulated or heated water systems can freeze if not properly maintained.

Solutions:

Clean waterers regularly: Check your waterers for blockages once a week, especially in the summer months when algae growth is more common. If using automatic systems, regularly inspect valves and hoses.

Invest in heated waterers: In colder climates, heated waterers or bases can prevent freezing and reduce your need to constantly check for ice.

Prevent leaks: Inspect hoses, seals, and connections regularly to catch small leaks before they become bigger problems. Have spare parts on hand for quick repairs.

Feeding System Problems

Automated feeders make it easy to keep chickens well-fed, but like any system, they can occasionally malfunction. Common feeding system issues and solutions include:

Feed clogging: Moisture or damp conditions can cause feed to clump and clog in automatic feeders, preventing the flow of feed.

Pest access: Rodents or wild birds can sometimes find ways to access feeders, leading to wasted feed and potential contamination.

Check for clogs regularly: Inspect your automatic feeder for blockages once a week. If feed is clumping, make sure the feed storage area is dry and consider switching to a different feed type that is less prone to moisture absorption.

Pest-proof feeders: Use rodent-proof feeders or install feeders designed to limit access by wild birds and other pests. Elevate feeders or use hanging feeders to keep rodents from climbing into them.

Ventilation and Temperature Control

Maintaining proper ventilation and temperature control is critical for your chickens' health. Too much heat or poor air circulation can cause stress, illness, and poor egg production.

Poor ventilation: Inadequate ventilation can lead to ammonia buildup from chicken waste, which can cause respiratory problems.

Overheating: In hot climates or during summer, chickens can easily overheat if the coop is not well-ventilated or shaded.

Cold drafts: In winter, drafts in the coop can cause discomfort or lead to frostbite in chickens, particularly on their combs and wattles.

Add adjustable vents: Make sure your coop has adjustable vents to allow for increased airflow in hot weather and decreased drafts in cold weather.

Shade and cooling systems: Use shade cloths, trees, or even solar-powered fans to keep the coop cool during hot months. Ensure there are shaded areas in the run as well.

Insulate but ventilate in winter: During winter, insulate the coop to reduce drafts, but do not seal it off entirely. Proper ventilation is still essential to prevent moisture buildup.

Egg Production Drops

Egg production can fluctuate naturally due to changes in weather, age, or diet, but prolonged drops in egg laying can indicate a problem.

Seasonal changes: As daylight hours decrease in fall and winter, hens will naturally lay fewer eggs or stop altogether.

Poor nutrition: A lack of key nutrients, especially protein and calcium, can reduce egg production or lead to thin-shelled eggs.

Stress: Changes in the flock, such as introducing new birds, or environmental stressors like predators, can cause hens to stop laying.

Supplement light in winter: Use timed artificial lighting to extend daylight hours to 14-16 hours per day, encouraging continued egg production in winter.

Check nutrition: Ensure that your chickens are getting a balanced diet rich in protein, calcium (like oyster shells), and fresh greens.

Reduce stress: Minimize disruptions in the flock, provide plenty of space, and make sure predators cannot access the coop or run.

Predator Problems

Predators such as raccoons, foxes, hawks, and even neighborhood dogs are a constant threat to backyard chickens. The most common issues include:

Predator breaches: Gaps or weaknesses in the fencing or coop structure can allow predators to enter.

Daytime predation: Hawks and other birds of prey may target free-range chickens during the day.

Secure fencing: Use hardware cloth instead of chicken wire for added strength, and make sure the fence extends underground to prevent digging predators from entering.

Automatic doors: Consider installing automatic coop doors that close at dusk and open at dawn to keep chickens safe overnight without needing manual intervention.

Netting for overhead protection: Use bird netting or poultry wire over the run to protect free-range chickens from hawks or other birds of prey.

Diagnosing Problems with Feeding, Watering, or Ventilation Systems

When a problem arises in your self-running system, the first step is diagnosing the issue correctly. Whether it is a feeding, watering, or ventilation problem, here's a quick troubleshooting guide to help you figure out the cause.

Feeding Issues: Is It the Feed or the System?

If your chickens aren't eating enough, or if the feeding system seems to be malfunctioning, consider the following:

Is the feed fresh? Old or spoiled feed can become moldy or stale, causing chickens to refuse it.

Is the feeder working? Check for clogs or blockages that may be preventing the flow of feed.

Are pests stealing the feed? Inspect the area for signs of rodents or wild birds that could be eating the feed before your chickens can get to it.

Watering Problems: Are Chickens Dehydrated?

If egg production drops or your chickens seem lethargic, dehydration could be the cause. Check your watering system for these issues:

Is the water clean? Dirty or algae-filled waterers can discourage chickens from drinking.

Is the system delivering water properly? Make sure there are no blockages, and check for leaks in hoses or pipes.

Is the water frozen? In cold weather, check waterers frequently to ensure they haven't frozen.

Ventilation Concerns: Stuffy or Drafty?

If your coop smells strongly of ammonia or if your chickens seem to have respiratory issues, ventilation could be the problem:

Is there enough airflow? Make sure vents are open and not blocked by bedding or debris.

Are drafts causing discomfort? In winter, seal off any cracks or gaps that could allow cold air to blow directly on your chickens, but leave vents open for air circulation.

Fine-Tuning Systems as You Learn What Works Best for Your Chickens

Every chicken coop is unique, and over time, you will learn what systems work best for your specific setup. Here's how to make adjustments and fine-tune your coop for maximum efficiency.

Experiment with Feeding Schedules and Amounts

Depending on your flock size and system, you may find that adjusting the amount or frequency of feeding works better for your setup.

Try different feeder types: Hanging feeders, gravity feeders, or treadle feeders all have advantages. Experiment with what works best for your chickens and environment.

Adjust feed quantities: Monitor how much feed your chickens consume daily and adjust accordingly. This will help prevent waste and ensure all chickens have access to food.

Improve Watering Efficiency

Watering systems can always be improved for ease of use and efficiency:

Automated waterers: Consider installing a larger water reservoir or a system that automatically refills from a rainwater catchment system.

Check flow rates: Make sure the water delivery system is fast enough to meet your flock's needs, especially in larger flocks.

Tweak Ventilation and Coop Design

Fine-tuning the coop's design and ventilation can help maintain a comfortable environment year-round.

Seasonal ventilation adjustments: In summer, keep windows or

vents wide open for maximum airflow. In winter, insulate the coop but maintain proper ventilation to prevent moisture buildup.

Raise the coop: If your coop does not have sufficient airflow from underneath, consider raising it on stilts or adding more ventilation at the base.

Long-Term Improvements: When and How to Upgrade or Expand

Over time, you may find that your coop needs upgrades or expansion as your flock grows or your goals change. Here are signs that it might be time for an upgrade and some strategies for making long-term improvements.

Signs it is Time to Upgrade or Expand

Overcrowding: If your chickens are fighting more frequently or egg production has dropped, it could be a sign that the coop or run is too small.

Increased maintenance: If you find yourself spending too much time on daily tasks, it may be time to invest in more automated systems or larger equipment.

New goals: If you are expanding your flock, raising meat birds, or hatching chicks, you may need to upgrade your setup to accommodate new needs.

Strategies for Upgrading and Expanding

Add a mobile coop: If your flock has outgrown its current space, consider adding a mobile chicken tractor to give your chickens access to fresh grazing areas.

Upgrade to solar-powered systems: Solar-powered lights, fans, or automatic doors can make your coop more efficient and reduce long-term energy costs.

Expand the run: If your chickens are free-ranging in a limited area, expanding the run with more fencing or using portable fencing can give them more space without much effort.

Continuous Improvement for a Self-Sustaining Coop

Even the best self-running chicken coops will occasionally need troubleshooting and fine-tuning, but with the right approach, you can maintain a smooth, efficient system. Whether you are dealing with a minor hiccup or considering long-term improvements, troubleshooting your coop effectively will keep your chickens healthy and happy with minimal disruption to your daily routine.

In the final chapter, we will summarize everything learned and discuss how to enjoy the freedom that comes with a self-sufficient chicken-keeping system.

Enjoying the Freedom of a Self-Sufficient Coop

After implementing the strategies and systems outlined in this book, you are well on your way to enjoying the benefits of a self-sufficient, low-maintenance chicken coop. The key to successful chicken keeping lies in setting up systems that work efficiently with minimal daily effort. In this conclusion, we'll recap the most important concepts, share final thoughts on balancing effort and reward, and encourage you to enjoy the freedom that comes with a well-managed, self-running coop.

Recap of Key Tips for a Low-Maintenance, Self-Running Chicken Coop

Throughout this book, we have explored various aspects of designing and maintaining a chicken coop that requires minimal effort while still providing excellent care for your flock. Here is a quick recap of the key takeaways from each section:

Choosing the Right Location and Setup

Location matters: Choose a spot for your coop that offers good drainage, plenty of sunlight, and protection from strong winds.

Size your coop properly: Ensure there is enough space inside the coop and in the run to avoid overcrowding, which can lead to stress and health problems.

Automate where possible: Install automatic feeders, waterers, and doors to reduce daily tasks.

Building a Coop for Durability and Ease of Maintenance

Use quality materials: Invest in weather-resistant, durable materials to ensure your coop stands the test of time and reduces the need for constant repairs.

Design for easy cleaning: Plan your coop layout with cleaning efficiency in mind—large access doors, removable roosts, and proper drainage will make cleanouts faster and easier.

Focus on ventilation and insulation: Proper ventilation ensures airflow while preventing drafts in colder months. A well-ventilated and insulated coop keeps your chickens comfortable year-round.

Simplifying Daily Routines

Efficient feeding and watering systems: Use gravity-fed feeders and automatic waterers to ensure your chickens always have access to food and water without daily intervention.

Natural light and airflow: Keep the coop well-lit with natural light during the day, and use solar-powered systems where appropriate to maintain consistency without relying on manual processes.

Regular health checks: Observe your chickens during feeding or egg collection times to catch any potential health issues early.

Sustainable Chicken Keeping

Composting chicken waste: Turn manure into valuable compost for your garden by balancing bedding and manure in a composting system.

Natural pest control: Incorporate herbs, plants, and dust baths to manage pests and reduce odors without the use of harsh chemicals.

Managing free-ranging: Allow chickens to forage naturally while ensuring their safety through proper fencing, shade, and predator protection.

Seasonal Maintenance

Spring cleaning: Perform a deep clean and check the coop for any repairs needed after winter.

Summer cooling: Provide shade, ventilation, and cool water to help chickens manage heat stress.

Fall prep: Get the coop ready for winter by adding extra bedding and securing insulation.

Winter care: Focus on maintaining airflow, preventing frozen water, and keeping chickens warm without overheating the coop.

Embracing the Simplicity of a Self-Running Coop

Once your self-running system is in place, chicken keeping becomes much more manageable and enjoyable. With systems in place to handle feeding, watering, ventilation, and pest control, you will find yourself spending less time on daily chores and more time enjoying the rewards of your flock.

Finding Balance Between Effort and Reward

A self-sustaining coop does not mean zero effort—there will always be some level of maintenance required. However, by automating as many processes as possible and scheduling regular seasonal tasks, you can significantly reduce the amount of hands-on work required. The key is finding a balance between effort and reward:

Daily tasks: Once your automated systems are running smoothly, daily tasks like egg collection and observation become opportunities to enjoy your flock rather than feeling like chores.

Seasonal upkeep: By breaking down larger tasks into manageable seasonal checklists, you prevent any one season from feeling overwhelming. These regular touchpoints help maintain the long-term health of your system.

Enjoying the Benefits of Self-Sufficiency

The true reward of a self-running chicken coop is the freedom and flexibility it offers. Whether you are away for a weekend or simply looking for a more relaxed pace of chicken keeping, a well-designed system takes care of the hard work for you. The benefits include:

Fresh eggs with minimal work: Enjoy the steady supply of homegrown eggs without spending hours a day on chicken care.

A thriving garden: Your chickens not only provide eggs, but also manure that can be composted and turned into rich fertilizer for your garden, creating a self-sustaining loop of food production.

Reduced feed costs: With natural foraging and pest control methods in place, you will spend less on feed and chemicals, making your operation more cost-effective and eco-friendly.

Final Thoughts on the Balance Between Effort and Reward in Chicken Keeping

Raising chickens can be a deeply rewarding experience, providing both practical benefits and personal satisfaction. By setting up a self-running chicken coop, you maximize those rewards while minimizing the effort required. With the right systems in place, you can enjoy all the joys of chicken keeping—fresh eggs, happy birds, and a healthy garden—without being tied down by constant maintenance.

In conclusion, whether you are new to chicken keeping or looking to improve your current setup, a self-sufficient, low-maintenance system can offer freedom and flexibility without sacrificing the health and productivity of your flock. With careful planning and a little fine-tuning, you can enjoy the best of both worlds: a thriving flock and the time to enjoy the fruits of your labor.

Simple Chicken Coop Setup Checklist for New Owners

1. Coop Design and Location

Select a location: Choose a well-drained area with plenty of sunlight and natural wind protection.

Size your coop: Allow at least 4 square feet per chicken inside the coop and 8-10 square feet per chicken in the run.

Build or buy a durable coop: Use weather-resistant materials like treated wood, hardware cloth, and metal roofing.

Plan for easy access: Include large doors or removable panels for cleaning and egg collection.

Provide proper ventilation: Install adjustable vents or windows to ensure airflow without drafts.

2. Bedding and Nesting Setup

Add bedding material: Use pine shavings, straw, or sand for bedding; ensure it is absorbent and easy to clean.

Set up nesting boxes: Provide one box for every 3-4 hens, lined with soft bedding or straw. Place in a quiet, dimly lit area.

Add roosting bars: Place roosts higher than nesting boxes to prevent chickens from sleeping in nests and soiling eggs.

3. Feeding and Watering Systems

Install automatic feeders: Set up a gravity-fed or treadle feeder that can hold several days' worth of food.

Install automatic waterers: Use a nipple waterer system or automatic waterer to ensure constant access to clean water.

Supplement with grit and calcium: Provide grit for digestion and oyster shells for calcium to strengthen eggshells.

4. Security and Predator Protection

Reinforce fencing: Use hardware cloth (not chicken wire) around the run and bury it at least 12 inches underground to prevent digging predators.

Install an automatic door: Set up a solar-powered automatic door that opens at dawn and closes at dusk.

Use locks on coop doors: Install secure latches to prevent raccoons or other predators from breaking in.

Protect from aerial predators: Add bird netting or a roof over the run to protect chickens from hawks and other birds of prey.

5. Lighting and Ventilation

Add natural light: Ensure there are windows or clear roof panels to provide plenty of natural light.

Install a timer for lights (optional): For consistent egg production, use a timer for artificial lighting in winter to provide 14-16 hours of light per day.

Set up a fan or solar-powered ventilation system: To keep air circulating in hot weather, consider a small fan or solar-powered fan.

6. Seasonal Preparations

Spring

Deep clean the coop: Remove all bedding, sanitize surfaces, and replace with fresh materials.

Inspect for winter damage: Check for leaks, drafts, or structural damage and repair as needed.

Prepare for pests: Use diatomaceous earth or natural pest repellents (like mint or lavender) to prevent mites and flies.

Summer

Maximize ventilation: Open windows, vents, or install shade cloths to keep the coop cool.

Provide shade and water: Ensure there is shaded space in the run and refill waterers with cool, fresh water.

Check bedding: Thin bedding in summer to improve airflow and reduce heat retention.

Fall

Prepare for winter: Add extra insulation to the coop (straw bales or extra bedding) and check for drafts.

Install heated waterers: Use heated waterers or heated bases to prevent freezing.

Start the deep litter method: Begin layering fresh bedding for warmth in winter.

Winter

Monitor ventilation: Keep vents open for air circulation but block drafts to prevent frostbite.

Check for frozen water: Ensure waterers aren't freezing; use heated options or change water frequently.

Increase feeding: Add extra grains (such as cracked corn) to increase calorie intake during cold months.

7. Free-Ranging and Pest Control

Allow free-ranging in safe areas: Use portable fencing or predator-proof areas for free-range time.

Create dust baths: Set up a dust bath area with dirt, sand, and diatomaceous earth to help chickens clean themselves and control pests.

Plant herbs: Grow lavender, rosemary, mint, and marigolds around the coop for natural pest control.

8. Egg Production and Health Monitoring

Collect eggs daily: Collect eggs at least once a day to avoid dirty or cracked eggs.

Provide oyster shells: Ensure chickens have access to oyster shells or another calcium source for strong eggshells.

Monitor chicken health: Check your chickens for signs of illness (lethargy, coughing, feather loss) during egg collection and feeding times.

9. Composting and Waste Management

Start a compost system: Compost chicken manure mixed with bedding to create rich fertilizer for your garden.

Use the deep litter method: Allow bedding to decompose inside the coop during winter, generating natural heat and compost.

Clean out the run regularly: Rake and remove any waste buildup in the run to prevent odor and flies.

Bonus Tips for a Self-Running System:

Automate where possible: Use automatic feeders, waterers, and coop doors to reduce daily chores.

Keep spare parts on hand: Store extra hoses, feeder parts, and small tools for quick fixes.

Observe flock behavior: Make sure your chickens are eating, drinking, and laying regularly—changes in behavior may indicate problems.

Authors Experience: Building Our Own Self-Running Chicken Coop

This book would not be complete without sharing our personal experience of building and maintaining a self-running chicken coop. While shared knowledge is essential, it is our hands-on experience over the years that truly brought the principles of this book to life. In this chapter, we will walk through the steps we took to build our coop, detailing the decisions we made based on our budget and what we learned along the way.

The Decision to Build Our Own Coop

One of the first decisions we faced was whether to buy a pre-made coop or build one ourselves. We ultimately chose to build our own because we strongly believed that, for the money, we could create something more durable and tailored to our needs. With the chicks still growing, we had a few months to complete the project, allowing us time to work on it at our own pace. The entire coop took around 100 hours to build, mostly utilizing one person.

Priorities and Planning

As emphasized throughout this book, the first priority was to find the right location. We chose the back of our yard, where there is a natural breeze and consistent airflow. The area is shaded by mature oak trees, which keeps the coop naturally cool during the hot summers. Living in a hot climate, temperature control was one of our top concerns, and we felt this location was ideal for keeping the chickens comfortable year-round.

Materials We Used

We decided to use wood, asphalt shingles, and hardware cloth. The hardware cloth was crucial for predator protection, given the presence of raccoons, possums, coyotes, hawks, and eagles in our area. We added a predator apron around the run to prevent digging animals from getting in.

We chose asphalt shingles for the roof because they offered great insulation and matched the aesthetics of our house. A well-designed coop should not only be functional but also complement the property.

Windows, Ventilation, and Accessibility

We installed windows all around the coop, but instead of opting for an automated door, we built a simple pulley system accessible from the outside, allowing us to manually close the door each night. If we are not home, it is no problem—the coop and run are fully predator-proof.

For cleaning, we made the coop's side windows double as doors that open widely. The lower sections also open, making it easy to clean the coop when necessary. Roost bars are removable, allowing for thorough cleaning. We installed a plexiglass front window to let in natural light, and we use the deep litter method inside the coop. The base of the coop is about a foot deep to allow the bedding to decompose naturally, providing warmth and reducing the need for frequent cleanouts.

Bedding and Pest Control

We chose industrial hemp for bedding inside the coop because it naturally repels flies and mosquitoes and can absorb about 10 times its weight in water. Furthermore, industrial hemp is a natural deodorizer. Hemp also keeps the coop dry and minimizes the need for frequent bedding changes. With hemp, we only need to replace the bedding every two years. When we used pine shavings, we had to replace them every six months; the hemp has been a big improvement.

The Run and Waste Management

In the run, we use construction sand because it is the best material we have found for breaking down chicken waste naturally. We also set up 5-gallon and 2-gallon waterers and feeders for redundancy, ensuring the chickens always have access to food and water.

The coop is designed to accommodate 8 to 10 chickens. The run is 13 x 8 feet, providing 108 square feet of space, while the coop itself is 8 x 4 feet, offering 32 square feet.

Elevated Coop Design

We raised the coop off the ground for three main reasons:

Avoiding humidity: This prevents the wood from rotting and keeps the floor dry.
Temperature control: The elevated design provides insulation from the heat and cold.
Shelter: The space under the coop offers a sheltered area for the chickens during bad weather.

Nesting Boxes and Egg Collection

We placed the nesting boxes on the side of the coop, allowing us to collect eggs easily from the outside. This setup also keeps the birds away from their waste. We wanted to avoid cleaning nesting boxes frequently, and this design keeps them cleaner. The hens rarely soil the nesting pads we use, and the eggs are always easy to collect.

Summer Cooling and Solar Power

To combat the summer heat, we installed a solar panel that powers a 12 volt fan to circulate air inside the coop on hot days. This helps oxygenate the space and keeps the air flowing, which has been a game-changer during heat waves.

Maintenance Routine

The run is cleaned about once every three years, while the coop is scooped weekly. We replace the bedding every two years, and all the old material is composted for our fruit trees, adding great value to our garden.

We also built the coop tall enough for us to walk inside without bending over, which makes cleaning much easier—especially as you get older!

Long-Term Success

Years later, our coop and run have accommodated seven healthy chickens. We initially started with three, but two years in, we added four more. They now produce 30-35 eggs per week, more than enough for our family. The chickens are happy, and so are we, as we spend more time enjoying their company and less time on constant cleaning.

Looking ahead, we plan to add a gutter water system and improve our automatic feeders. We have also opted not to use lights inside the coop, as the hens already produce more than enough eggs for us, and we believe they deserve a natural break from laying during the winter months.

Final Thoughts on Building a Self-Running Coop

Over the years, friends and family have asked us about our self-running coop. At first, it seemed like a daunting task to create something that required so little daily work. But after years of research and hands-on experience, we have built a system that truly works. The results have been amazing. Our coop runs efficiently, and we now enjoy chicken keeping in a way that's hard to imagine living without.

The beauty of a self-running system isn't just about reducing chores—it is about enhancing the quality time spent with your flock and producing high-quality eggs, and have happy birds. We hope you found value in this book and that you, too, can create a system that brings you as much joy as ours does for us. Your journey may vary, but it is entirely possible to have a self-sufficient chicken coop that works for both you and your chickens.

Thank you for reading, and we wish you all the best in your chicken-keeping adventures!

Biographies

Dr. Julia Diakakis

Dr. Julia Diakakis, a highly respected and decorated educator and businesswoman, boasts an impressive career spanning over 35 years. Born into a military family in Bremerhaven, Germany, Dr. Diakakis was raised throughout Europe providing her a diverse upbringing which laid the foundation for a versatile professional journey.

A proud graduate of Ohio University, she holds a B.S. in Education with a minor in English. Her certifications span various subjects including English, Math, History, Social Science, Earth Space Science, and Elementary Education with endorsements in Reading and English Speakers of Other Languages (ESOL). Her passion for education led her to different roles in educational institutions ranging from elementary to high school levels, earning her recognition as one of Florida's most esteemed educators. Furthering her passion for education, she has served on dissertation committees as the content expert for doctoral students.

Parallel to her educational career, Dr. Diakakis has been maintaining an IT business since 1994 and served the Brevard School Board as an IT Specialist for seven years. This combination of education and technology expertise uniquely positions her at the intersection of these two critical fields.

Dr. Diakakis further expanded her skill set by obtaining a master's degree in educational administration, leading to her principal certification and an administrative role for about 15 years. She didn't stop there; her boundless passion for education, business, and information technology propelled her to pursue a Doctorate in Organizational Leadership and Development.

A member of prestigious societies such as Alpha Chi Honor Society, Kappa Delta Pi Honor Society, and National Society of School Administrators, Dr. Diakakis has made significant contributions to developing computer lab programs for Brevard Public Schools in Florida. She also pioneered the first fully operational Discourse Study Computer classroom in Florida. Her extensive work with numerous private companies throughout her career testifies to her business acumen.

Dr. Diakakis' unique blend of educational knowledge and business insights makes her a formidable force in her field. Her life and career serve as an inspiration for many who aspire to make a difference in the realms of education and business.

Contact

E-mail: authors@diakakis.us

Websites:

https://diakakis.us

https://BrevardHost.net

https://alphaomegacom.net

Saki Diakakis

Born in Athens, Greece, Saki Diakakis embarked on his journey with a classic high school education followed by military service as an Amphibian Green Beret. In 1989, Saki immigrated to the United States to nurture his passion for aviation and began training as a private pilot, initially aiming for a career as a commercial pilot. However, he found himself drawn towards the burgeoning field of computers, networks, and software development. Saki returned to academia, this time immersing himself in computer science, and subsequently ventured into diverse IT industries.

From the early 1990s, his career path led him to work with local city government bodies as a Network Systems Analyst and Software Developer. His professional portfolio expanded across multiple sectors including county, city, and state governments, in areas of advertising, space subcontractors, home builders, and water treatment plants, among others. During his time with city government he designed and built the TV station for the City of Cocoa Beach, marking a significant milestone in his career.

During a nine-year stint in the advertising industry, Saki managed central networks and developed integral systems for the national yellow page company. As a consultant, he contributed to data analysis and application development. Over the years, he has developed an array of commercial applications encompassing document management systems, CRM, billing, building department permits, fleet management, and more. Adding another feather to the cap, he pioneered one of the first surveillance wireless undercover applications in the United States.

For the past 12 years, Saki has been associated with a local Managed Services Provider, specializing in computer networks, telephony, and cybersecurity applications. His services have benefited hundreds of businesses on local, federal, and international levels.

Saki's extensive formal education and practical experience have equipped him with knowledge that many businesses and stakeholders consider invaluable for their success. He is committed to leveraging this expertise to continue the contribution to the field of technology and business.

Contact

E-mail: authors@diakakis.us

Websites:

https://diakakis.us

https://BrevardHost.net

https://alphaomegacom.net

Notes:

Notes:

Made in United States
North Haven, CT
04 November 2026

21736091R00108